PROBABILITY:
AN
INTRODUCTORY
COURSE

PROBABILITY: AN INTRODUCTORY COURSE

BY

NORMAN R. DRAPER
UNIVERSITY OF WISCONSIN

AND

WILLARD E. LAWRENCE
MARQUETTE UNIVERSITY

MARKHAM PUBLISHING COMPANY
CHICAGO

Preface

The study of probability was first brought to the serious attention of mathematicians as a result of interest in the odds of winning games of chance. It is recorded that the seventeenth-century French gambler, the Chevalier de Méré, sought the advice of the famous mathematician Pascal about a problem concerning a certain dice game: How could one determine the probability each player has of winning the game at any given stage? Pascal and Fermat, and probably others, discussed this problem and the solution was found. Subsequently, a mathematical theory of probability gradually emerged in the published works of prominent mathematicians and other scientists.

At the present time, probability occupies a fundamental and respected role in the scientific world. As a branch of pure mathematics, it has been developed on a rigorous, axiomatic basis and is the subject of much current study and research. As a tool in science, business, and industry, it has many applications; many phenomena formerly regarded as deterministic in nature have been reinvestigated from a probabilistic viewpoint, which has often been found to provide results more consistent with actual experience. The study of probability is also a stepping stone to the study of statistics, which has applications in every scientific field.

This book is intended for readers who wish to acquire some of the basic notions of probability theory but who have not taken a course in calculus. Only a facility with high school algebra methods is needed to begin.

As a course text, the book is suitable for college students at all levels, but we believe that well-prepared high school students will also find the book readable and will (with guidance) be able to master the material presented. We hope that other readers, who may be interested in probability but who are not students in a formal program, will find the book helpful and interesting as well.

In Chapters 1 and 2 we present background material necessary for the study of probability theory. The real world in which the laws of probability operate and the ideal world of mathematical models are explored and brought together. Several methods of assigning probabilities are discussed and illustrated. Sophisticated methods of counting which simplify both notation and arithmetic are explained, and an introduction to the language of sets and the rudiments of set theory is provided.

The study of probability theory proper begins with the definition of a sample space and a probability function in Chapter 3. The special case of equally likely outcomes is used to introduce and to illustrate some of the laws and theorems of probability.

The study of several important distributions in Chapters 4, 5, and 6 is preceded by the idea of a random variable. Although the book is concerned primarily with discrete random variables, the uniformly distributed random variable is mentioned to point out the similarities and differences between discrete and continuous random variables. A treatment of Chebyshev's inequality is given.

In Chapters 7 and 8 the notion of several random variables defined on the same sample space is explored. Regression and correlation are treated and Chapter 8 ends with a discussion of the central limit theorem.

Discrete Markov processes are dealt with at length in Chapter 9, and in Chapter 10 is sketched the basic problem of decision-making using expectations. These two chapters are intended as illustrations of the practical value of probability.

When used as a class text, the book contains sufficient material for a one-semester introductory course in probability. The material has also been used successfully as a one-semester course in finite mathematics for liberal arts freshmen. The main objective here was to give the student a course in modern mathematics with some mathematical rigor and practical application; a short course in probability theory does this admirably.

Most of the students who use this book will be already familiar with dice and playing cards. Those who are not should obtain a pair of dice and a pack of playing cards immediately, because familiarity is assumed in a number of the examples and exercises.

Section 1.5 may be omitted by readers with a knowledge of permutations and combinations, and Chapter 2 may be omitted by those familiar with the notation and algebra of sets. Some possible courses of various durations (exclusive of examinations) are as follows:

Chapters 1–10	48 hours
Chapters 1–8	38 hours
Chapters 1–5, 8	27 hours
Chapters 1–5, 8 (omitting Sections 3.5, 3.7, 5.2, and 5.4)	23 hours

For students who do not need the material on permutations, combinations, and sets, the following courses are feasible:

Chapters 1, 3–10 (omitting Section 1.5)	43 hours
Chapters 1, 3–8 (omitting Section 1.5)	33 hours
Chapters 1, 3–5, 8 (omitting Section 1.5)	22 hours

We are grateful to Professor William Feller, Professor Emanuel Parzen, and their respective publishers, John Wiley & Sons, Inc., and Holden-Day, Inc., for permission to use material appearing in Chapter 9.

We are also grateful to the following typists: Susan D. Anderson (now Costello), Mary Ann Clark, and Wanda R. Gray, all of Madison, and Susan M. Berger of Milwaukee, for their careful work throughout the preparation of this book.

In addition, we should like to thank Mary Esser, office supervisor of the Statistics Department, University of Wisconsin, for her never-failing common sense and patience, and for her help in numerous ways.

We shall be glad to hear about any misprints or errors that readers may find.

NORMAN R. DRAPER
WILLARD E. LAWRENCE

Contents

CHAPTER 1
Prediction and Probability

1.1. Predictions and Uncertainty

Have you ever made a prediction? Of course you have! Perhaps you have predicted that:

1. The Cardinals will win the World Series.
2. It will snow on Christmas Eve.
3. Skirts will get longer next year.
4. The Republican candidate will win the next election.
5. A tossed coin will come down heads.
6. You will get a B in a certain mathematics course.
7. England will become the fifty-first state of the union.

When you make predictions such as these you may be taking a wild guess or you may "know something." For example, if the coin you are about to toss has heads on both sides, your prediction of heads is a sure thing. The event you predict is certain to occur. Suppose that in your mathematics course you have already made B grades on the first three examinations and feel you have done work at your usual level in the final. Your prediction of a B grade overall is not a complete certainty but you will probably feel extremely confident about it. However, if you toss an ordinary coin that has a head on one side and a tail on the other, and you predict that it will fall heads, you may feel that you will just as likely be wrong as right in your prediction.

How do we indicate how much confidence we have in our predictions? One way is to think of what odds we would be prepared to take if we wanted to bet $1 that our prediction would be correct. On a sure thing we would be prepared to give enormous odds, perhaps 1 to 1,000,000. For example, if we had predicted that a two-headed penny would fall heads we would be prepared to win $1 if we were right and to lose $1,000,000 if it fell tails, because it is impossible that it *will* fall tails. On the other hand, if the coin were an ordinary one, we would probably feel that odds of 1-1 (that is, "evens") were appropriate. In other situations we might wish to take odds. For example, if we took odds of 10-1 we would win $10 if we were correct but lose only $1 if we were wrong.

Now, naturally, if two opponents are about to toss a coin and both *know* that the coin is double-headed, there will be no betting. The result of the toss is completely certain and no element of uncertainty is present. Bets are made only in situations where uncertainty exists, that is, on phenomena that have more than one possible outcome, or at least *appear* to have more than one possible outcome to at least one of the persons involved.

What do we mean by this last remark? Well, suppose we have an ordinary coin with one side heads and the other side tails. If we predict the result of a toss as heads and want to bet on it, we may find someone willing to predict tails and bet on that, because the outcome appears uncertain to him. Suppose, however, that by long hours of practice we have been able to control the tossing so that the coin comes down heads on every toss. Because of this control over the outcome, we now have a sure thing, even though our adversary is unaware of it.

Perhaps, even though we have practiced many hours, we have not been able to control the toss completely. However, we may have partial control, and perhaps can make the coin come down heads about three-fourths of the time. The result of the toss is now neither purely haphazard nor perfectly under control but somewhere in between. There is still *some* uncertainty about the outcome, of course, but if our betting opponent knows the true situation, he will certainly not bet even money. Because he knows that we will probably win about three-fourths of all the tosses, he will want some suitably adjusted odds. (If he does not know the true situation, he will no doubt begin to suspect it when he finds himself on the losing end of the bet most of the time.)

The point we wish to make by the examples above is this: A phenomenon may have only one outcome (which is a sure thing) or there may be several possible outcomes. In the second case, if we wish to make a bet on one of these outcomes, we shall want to know the chances with which the various outcomes will occur. In this way we can perhaps make predictions with some measure of success.

The study of probability is concerned entirely with occurrences whose outcomes are uncertain to some degree. (As a special case we shall even be able to fit sure things into our formulation.)

What types of occurrences involve uncertainties? We often work with simple examples such as the result of a coin toss, or the throw of a die, or the spin of a pointer, or the selection of a card from a deck. There are, however, many other happenings in our everyday lives that have an element of uncertainty about their outcomes. The familiar quote states that only two things are certain, death and taxes, but even these can be subjects of a probability study. Death is certain, but the time, place, cause, and other circumstances are uncertain. Taxes seem to be with us for the foreseeable future, but their levels and their effects are uncertain.

1.2. Random Phenomena

When a phenomenon has an uncertain outcome, we shall call it a *random phenomenon*. The phenomenon may be a happening, a state or condition, a planned experiment, a survey, a study, an act of nature, an accident, or anything else that has an outcome that can be observed. (By observed we may mean several things: The outcome may be seen, heard, tasted, felt, smelled, or measured in a specified manner.) The outcome of interest may be just one attribute of the whole phenomenon. For example, the phenomenon may be the selection of a person off the street and the attribute may be his weight, height, or age. In other cases the outcome to be observed or measured may be a combination of several attributes. In the post office, for example, the size of a package is assessed by adding together its length and its girth.

To make our description of a random phenomenon a little more precise, we condense it into the following statement, which may serve as a definition: *A random phenomenon is one about which there is some uncertainty because it has more than one possible outcome.* If it is to be interesting, the phenomenon should happen more than once, and preferably we should be able to observe it repeatedly, or there will not be much point in studying it. Also, we should be able to identify the set of all the possible outcomes of the phenomenon, and, furthermore, there should be no regular pattern or sequence of actual outcomes, or the uncertainty vanishes.

Consider the result of the toss of an ordinary six-sided die. Is this a random phenomenon? If the die is not biased in any way and the tossing is impartial, then the outcome of a single toss (the number of dots showing on the upper face) is uncertain, so the answer is yes. All we know for sure is that the outcome is an integer between one and six inclusive. The phenomenon can be observed repeatedly, as often as the die is tossed.

If a phenomenon has only one outcome which can be determined precisely before its observation, it is said to be a *deterministic phenomenon*. These are actually less common in real life than one would at first think. For example, we all "know" that, at a temperature of 32° Fahrenheit, water becomes ice. If the temperature falls to 32°F, is the formation of ice certain? Yes, if the water, and all the local conditions, are under strict control, and the measuring instruments are accurate. In practice, if water is cooled to 32°F we may not be able to say whether ice will form. The possible inaccuracy in measuring the temperature, the purity of the water, the atmospheric pressure, and other factors could all affect the ice formation. Given enough information about all the relevant factors, we could perhaps determine the probability that ice will form. However, if the factors are completely uncontrolled, the phenomenon is one whose outcome is not certain.

In general, it is almost impossible to fix all the factors affecting the outcome of a phenomenon. One reason for this is that, in many cases, not all factors are known or even suspected. Even if they *are* known, it is unlikely that they will all be controllable. Another reason is that some factors are quantitative; that is, the factor is a continuous measurement such as the weight, size, or temperature. Because of limitations of measuring devices and of errors in reading the scale, every measurement is only an approximation to some ideal value.

When the factors that influence a phenomenon cannot be controlled, the situation cannot be duplicated precisely enough to ensure exactly the same outcome every time the phenomenon occurs. The phenomenon is then not deterministic, but random.

Games of chance, such as tossing coins or dice, provide simple, well-known examples of the properties of random phenomena. However, random phenomena exist in many fields of study. Students of physical science, social science, life science, engineering, business, journalism, and other fields will find many examples in their own fields. They will also find the theory of probability a valuable tool for examining the random phenomena in these fields.

Exercises

1. Are the phenomena involved in the following situations random or deterministic?

    (a) A coin is tossed to determine which team in a football game gets its choice of receiving or kicking off.

    (b) When 111 is divided by 37 on a desk calculator, the answer 3 comes up.

    (c) Four people are dealt bridge hands; South's hand contains 13 cards; 5 of these are clubs, 3 are diamonds, 3 are hearts, and 2 are spades.

    (d) Names of teams are drawn to determine pairings in a basketball tournament.

    (e) A girl buys a packet of bobby pins in Woolworth's; there are 20 pins in the packet.

    (f) The number of incoming telephone calls to Marquette University is recorded for every 1-minute period during selected hours.

2. Write down two examples of random phenomena and two of deterministic phenomena in your own life.

3. Give examples of the use of probability theory that you have heard or read about in fields such as the physical and social sciences, business, engineering, and medicine.

4. A coin is tossed. If the coin is

    (a) balanced (that is, fair)   (c) two-headed

    (b) unbalanced (that is, unfair)   (d) Portuguese

is the toss a random or a deterministic phenomenon?

5. Say why each of the following is, or is not, a random phenomenon:

    (a) The occurrence of the digit 6 as the last digit of a telephone number in a large directory.

    (b) The occurrence of the letter M as the last initial of a person at a large gathering of people.

    (c) The occurrence of a Chevrolet arriving at a traffic signal on a busy highway.

    (d) The occurrence of red at a traffic signal.

    (e) The appearance of a large policeman as you run a red traffic signal.

1.3. Mathematical Models

In all the experimental disciplines (for example, physics, chemistry, or biology), mathematical models are very common. Whenever we postulate a specific mathematical model that does not contain any random elements, we are postulating what would happen in ideal circumstances. If, for example, all factors that affect the outcome of an experiment can be realized without error, the outcome would be predicted by our mathematical formula. Moreover, our prediction would be the same every time the experiment was performed with the factors at the same levels. To take a simple example: Suppose we postulate that an outcome z is affected by factors x and y according to the model $z = 2x - 3y$; then we should always predict the outcome to be 4 when $x = 5$ and $y = 2$.

In an actual performance of such an experiment, however, we would not expect to obtain a result of exactly 4, for several reasons:

1. There may be factors other than those represented by x and y which affect the outcome.

2. The physical settings of the values x and y may be subject to error.

3. The measurement of the outcome, z, may also be subject to error.

Two people trying to set values of x and y are quite likely to obtain different actual values and hence different values of z. Moreover, two people measuring the outcome z for identical settings of x and y are likely to report slightly different results.

The mathematical model, then, is an idealization of a physical situation. Simple models have already been used by the student in solving "word" or "story" problems in algebra, and problems in physics. For example, the mathematical model for the position of a freely falling body with zero initial velocity and at zero position initially is $s = \frac{1}{2}gt^2$, where g is the acceleration in feet per second per second due to gravity, t the elapsed time in seconds, and s the distance fallen in feet. This is a deterministic model. It supposes that values of g and t are precisely given. If they are, the position of the falling body at any time is a single determined value.

Mathematical models can also be used to describe random phenomena. A random phenomenon has several possible outcomes and so a model must provide the means of evaluating the probabilities that the various outcomes will occur. For this reason it is called a probability model (as opposed to a deterministic model, which gives the outcome that is *certain* to occur if the conditions involved are satisfied).

To set up a probability model it is necessary to know all the possible outcomes of the random phenomenon and to have either theoretical or empirical justification for the formulation adopted. We shall consider a number of important models in detail later. For the moment, however, we present a few examples of probability models without providing justification for them.

Example 1. Suppose a balanced or fair coin is tossed three times. The following mathematical model provides the probabilities of obtaining 0, 1, 2, or 3 heads:

$$p_k = \text{probability of obtaining } k \text{ heads in three tosses}$$

$$= \binom{3}{k}\left(\frac{1}{2}\right)^k\left(\frac{1}{2}\right)^{3-k} = \binom{3}{k}\left(\frac{1}{8}\right),$$

where

$$\binom{3}{k} = \frac{3!}{k!(3-k)!}$$

and where $k! = 1 \cdot 2 \cdot 3 \cdots (k-1) \cdot k$ (see Section 1.5). The derivation of this formula will be considered when the binomial distribution is discussed later.

The probability model above gives the probability of obtaining $k = 0$, 1, 2, or 3 heads in three tosses of a coin, provided that the probability of getting a head on a single toss is $\frac{1}{2}$. The model may be represented in several ways. Above it is given as a formula. We can evaluate p_k for the various values of k and represent it as a table:

k	0	1	2	3
p_k	$\frac{1}{8}$	$\frac{3}{8}$	$\frac{3}{8}$	$\frac{1}{8}$

Thus there is a probability of $\frac{3}{8}$ that one head (or two heads) will occur in three tosses of a fair coin, and a probability of $\frac{1}{8}$ that no heads (or three heads) will occur in three tosses. We can also represent the probabilities as "spikes" on a figure. Each spike in Figure 1.1 represents the corresponding value of p_k in the table above. Thus the spikes at $k = 1$ and $k = 2$ are equal and are three times as high as the spikes at $k = 0$ and $k = 3$, which are also equal. Figure 1.2 shows another way of representing the same probability model, with "bars" instead of spikes. Corresponding heights are the same, however. Note that $p_0 + p_1 + p_2 + p_3 = 1$. A probability of one means "certainty," and this confirms the fact that when three coins are tossed, 0, 1, 2, or 3 heads must occur and there are no other possibilities.

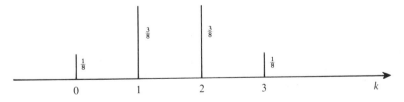

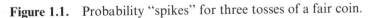

Figure 1.1. Probability "spikes" for three tosses of a fair coin.

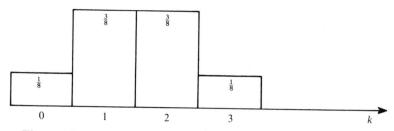

Figure 1.2.   Probability "bars" for three tosses of a fair coin.

Example 2.   The probability that the thirteenth day of a month selected at random will fall on each of the seven days of the week is

Day of week	Sun	Mon	Tues	Wed	Thurs	Fri	Sat
Probability	$\frac{687}{4800}$	$\frac{685}{4800}$	$\frac{685}{4800}$	$\frac{687}{4800}$	$\frac{684}{4800}$	$\frac{688}{4800}$	$\frac{684}{4800}$

Example 3.   Suppose there are x people in a room. The probability $p(x)$ that two or more of these persons have the same birthday is given by the following table:

x	4	8	12	16	20	22	23
$p(x)$	0.016	0.074	0.167	0.284	0.411	0.476	0.507

x	24	28	32	40	48	64
$p(x)$	0.538	0.654	0.753	0.891	0.961	0.997

Thus, for example, if there are 23 people in the room the chances are better than even that two people will have the same birthday. When 64 people are present, it is almost certain that two of them will have the same birthday.

Example 4.   If there are x students in class and a teacher returns x graded test papers randomly, the probability that each student will get his own paper is $1/x!$

Exercises

1. (a) Following the example in the text for the probability of getting 0, 1, 2, or 3 heads in three tosses of a balanced coin, write down the mathematical model (that is, the formula) that will provide the probabilities of getting 0, 1, 2, 3, 4, or 5 tails in five tosses of a balanced coin, and calculate the probabilities. What is the sum of the probabilities? Why?

(b) Assume that the formula in the text can be used as the formula for the probability that, of three children born, exactly k will be boys. Find the probability that all three will be girls. What basic assumption are we making here?

(c) Write what you think would be a suitable formula that gives the probability of k boys for the case when six children are born. *Guess* the probability that three will be boys, and then work out the actual probability.

(d) Write what you think would be a suitable probability model that will provide the probability of getting $0, 1, 2, \ldots, 100$ heads in 100 tosses of a balanced coin.

2. If 10 school children put their lunch bags into a large box and at lunch time each takes a lunch bag at random from the box, find the probability that each child will have his own lunch. (Use the results of Example 3.)

1.4. Methods of Assigning Probabilities

Suppose a random phenomenon has several possible outcomes and we conduct a single experiment, or observe the phenomenon once, whichever is physically appropriate. (We can say that we have conducted a *trial* in either case.) What is the probability of a particular outcome? In the most general sense it is a measure of the chance that the particular outcome will be observed. This measure, which we can call a *probability measure*, is a number assigned to the outcome to indicate this chance. By convention, the number is nonnegative, real, and not greater than one. In other words, the probability of any particular outcome is a real number between zero and one inclusive which expresses the chance that this outcome will be observed. A value near zero indicates small chance and a value near one indicates large chance.

How is such a number chosen? There are three general methods: classical, relative frequency, and subjective.

THE CLASSICAL METHOD. Here, probabilities are assigned theoretically according to the following rule: If a random phenomenon has N possible outcomes all of which are "equally likely" to occur, the probability that a particular outcome will be observed is $1/N$. Furthermore, if we select a group of n outcomes, the probability that any one of the specified group of n outcomes will occur is n/N.

The equally likely part of the classical method is an important one and often, in examples, implausible results are obtained when it is not taken into account. For example, if we toss two coins, we can categorize the possible outcomes in several ways, two of which are:

1. 0 heads, 1 head, 2 heads.
2. (T, T), (H, T), (T, H), (H, H).

Suppose both coins are fair (or balanced). Then the classical method cannot be applied in the first case because the three outcomes are not, in fact, equally likely. Thus to assign them probabilities $\frac{1}{3}, \frac{1}{3}$, and $\frac{1}{3}$ would be incorrect. However, the four outcomes in the second case *are* equally likely and to assign them probabilities of $\frac{1}{4}$ each would be perfectly correct. Note that this implies that "1 head" is, in fact, *twice* as likely as either 0 or 2 heads.

Whenever (apparently proper) use of the classical method gives answers that seem to contradict intuition and experience, the assumption that certain outcomes are equally likely needs to be reexamined.

Example 1. A fair coin is tossed. The classical method assigns the probability of $\frac{1}{2}$ to the outcome "heads," for there are only two ($N = 2$) possible outcomes, one of which is "heads."

Example 2. A fair die is tossed. There are six equally likely outcomes, so the probability that the number of dots showing will be two is $\frac{1}{6}$. The probability that the die will show an even number of dots is $\frac{3}{6}$, for three ($n = 3$) of the six ($N = 6$) possible outcomes are even numbers.

Example 3. If one fuse is selected at random from a box containing 20 fuses of which 5 are known to be burned out, the probability of selecting a good fuse is $\frac{15}{20}$.

Example 4. If a basketball player is selected at random from a group of 25 players of whom 3 are centers, the probability of getting a center is $\frac{3}{25}$.

Example 5. If a card is selected at random from a bridge deck, the probability of getting an ace is $\frac{1}{13}$, and the probability of getting a spade is $\frac{1}{4}$.

RELATIVE FREQUENCY METHOD. Here probabilities are assigned through experimental, or empirical, means. As an example, consider the experiment of tossing a coin. Suppose that 60 heads are observed in the first 100 tosses (or trials). The proportion of heads is $\frac{60}{100}$, or 0.60. This number is the relative frequency of heads. Suppose that 55 heads are observed in the next 100 tosses, giving a total of 115 heads in 200 tosses. The relative frequency of heads is now $\frac{115}{200}$, or 0.575. If, as the number of repetitions of the 100 tosses gets larger and larger, the relative frequency of heads fluctuates slightly about a certain value, say 0.583, but seems to be steadying down at

0.583, then this value is taken as the probability that this particular coin will turn up heads on a single toss. (Calculating the relative frequency of heads after each set of 100 tosses is purely for convenience; strictly speaking, the calculation should be made after every toss.)

To take another example, suppose we wish to predict the chance that a particular outcome will occur when we throw a die. We might do something like the following: Toss the die 120 times and keep a record of the frequencies of ones, twos, threes, fours, fives, and sixes that actually occur. These frequencies divided by 120 give the relative frequencies. (Each number *should*, if the die is fair, occur about 20 times in fact—about one-sixth of the total number of tosses.) If the die is tossed another 120 times, we can get the relative frequencies for the whole 240 tosses. (These should still be about one-sixth for each number for a fair die. In fact, we should expect them to be closer to one-sixth than after the first tosses, although this need not be the case.) As this experiment is repeated and the number of tosses increases, we would expect the relative frequencies of occurrence of each of the six possible outcomes to settle down close to some fixed values (which, in fact, should be one-sixth for a fair die). The relative frequencies are the numbers we should use for our predictions of the chances of the particular outcomes.

Consider the following experiment performed to determine the probability of getting a six when tossing a certain die. The die was tossed 800 times, with the results given in Table 1.1.

Table 1.1. Frequency of a Six in 800 Throws of a Die, Split into Groups of 200 Throws Each

(1)	(2)	(3)	(4)	(5)	(6)	(7)	(8)
						Relative	
				Cumulative totals of:		frequencies of:	
		No. of results other		Results Throws		Sixes	Results other than six
Group No.	No. of sixes	than six	Sixes	other than six	[col. (4) +col. (5)]	[col. (4) ÷col.(6)]	[col. (5) ÷col.(6)]
1	3	17	3	17	20	0.1500	0.8500
2	3	17	6	34	40	0.1500	0.8500
3	3	17	9	51	60	0.1500	0.8500
4	2	18	11	69	80	0.1375	0.8625
5	2	18	13	87	100	0.1300	0.8700
6	4	16	17	103	120	0.1417	0.8583
7	7	13	24	116	140	0.1714	0.8286

Table 1.1.—*continued*

(1)	(2)	(3)	(4)	(5)	(6)	(7)	(8)
				Cumulative totals of:		Relative frequencies of:	
Group No.	No. of sixes	No. of results other than six	Sixes	Results other than six	Throws [col. (4) + col. (5)]	Sixes [col. (4) ÷col.(6)]	Results other than six [col. (5) ÷col.(6)]
8	3	17	27	133	160	0.1688	0.8312
9	2	18	29	151	180	0.1611	0.8389
10	1	19	30	170	200	0.1500	0.8500
11	6	14	36	184	220	0.1636	0.8364
12	3	17	39	201	240	0.1625	0.8375
13	3	17	42	218	260	0.1615	0.8385
14	2	18	44	236	280	0.1571	0.8429
15	5	15	49	251	300	0.1633	0.8367
16	4	16	53	267	320	0.1656	0.8344
17	3	17	56	284	340	0.1647	0.8353
18	3	17	59	301	360	0.1639	0.8331
19	5	15	64	316	380	0.1684	0.8316
20	3	17	67	333	400	0.1675	0.8325
21	3	17	70	350	420	0.1667	0.8333
22	4	16	74	366	440	0.1682	0.8318
23	2	18	76	384	460	0.1652	0.8348
24	2	18	78	402	480	0.1625	0.8375
25	7	13	85	415	500	0.1700	0.8300
26	1	19	86	434	520	0.1654	0.8346
27	1	19	87	453	540	0.1611	0.8389
28	4	16	91	469	560	0.1625	0.8375
29	7	13	98	482	580	0.1690	0.8310
30	2	18	100	500	600	0.1667	0.8333
31	3	17	103	517	620	0.1661	0.8339
32	4	16	107	533	640	0.1672	0.8328
33	3	17	110	550	660	0.1667	0.8333
34	5	15	115	565	680	0.1691	0.8309
35	2	18	117	583	700	0.1671	0.8329
36	1	19	118	602	720	0.1639	0.8361
37	4	16	122	618	740	0.1649	0.8351
38	4	16	126	634	760	0.1658	0.8342
39	3	17	129	651	780	0.1654	0.8346
40	4	16	133	667	800	0.1662	0.8338

The relative frequency for a six is seen to vary considerably at the beginning of the experiment, but it steadies down toward the end and appears to fluctuate about roughly the theoretical value $\frac{1}{6}$. Similarly, the relative frequency for "other than six" steadies down and appears to fluctuate about the value $\frac{5}{6}$. (The probabilities of getting the other five numbers on the die could, of course, be obtained in a similar manner.)

Comments. The conventions established earlier can now be partially justified, for suppose we have a random phenomenon with several possible outcomes. Whether we use the classical method or the relative frequency method, the number assigned as the probability of a particular outcome is the ratio of two integers, the denominator being positive and the numerator being either positive or zero. Thus the probability is a nonnegative real number. Furthermore, the ratio can never be greater than one. (Why?) Also, the sum of the probabilities assigned to all possible outcomes of a random experiment cannot be greater than one, and is in fact exactly equal to one. (Why?)

SUBJECTIVE PROBABILITY. The types of probability we have talked about so far have been objective, that is, independent of what is personal or private in an individual's feelings. However, probability may also be subjective, that is, peculiar to a particular individual, and modified by that individual's bias and personal limitations, in other words, dependent upon the individual observing the phenomenon. If the probability of an outcome is determined by using all the possible information and facts about the phenomenon, the probability will be an objective one. However, if the probability is to be determined by an individual by considering only the limited information he actually has about the phenomenon, the probability so determined is subjective. Two people may have completely different amounts of information about the phenomenon. We can thus see that, although in theory a probability may be considered as an objective one, in practice it may well actually be a subjective one, if full information is not available.

The nature of the particular phenomenon has much to do with whether the probability is subjective or objective. For example, if a coin is tossed a great many times and the relative frequency of heads is calculated to be 0.524, everyone aware of the results would quote the probability of the outcome heads as 0.524. Suppose, however, that the coin is not tossed but, instead, several persons are given it in turn to examine briefly and are then asked to give the probability of heads. The probability given in each case would depend upon what each person observed in his examination, and this is subjective.

Two people may well have different degrees of confidence in the prospect of rain for tomorrow. The degree of confidence would depend upon several pieces of information as well as upon a person's experience, training, and so on. One person may have studied meteorology and be aware that, with certain combinations of wind direction, pressure, temperature, and cloud formation, the probability of rain is very high; another person may not know this. One person may have heard a weather forecast; the other person may not have. One person may have joints that ache when rain is imminent; and the other person may have no such aid.

The degree of confidence (properly expressed as a fraction or decimal lying between zero and one inclusive) could be taken as the probability of rain. Each person would have his own probability, and these might very well be different. Individual numerical probability values could be obtained by a method such as the following. A person is given his choice of two opportunities:

1. He will win $1 if it rains tomorrow, or

2. he will win $1 if a number, picked at random from slips of paper numbered 1 to 100, is greater than 50.

If he selects the first choice, his subjective probability of rain is greater than $\frac{1}{2}$, because the probability of getting a number greater than 50 is $\frac{1}{2}$, and he obviously believes that the chance of rain is greater than the chance of getting a number greater than 50. If he still picks the first choice when the number is 25 instead of 50, his probability of rain is greater than $\frac{3}{4}$, for the probability of getting a number greater than 25 is $\frac{3}{4}$. By continuing in this manner, a value of his probability of rain is determined. (We say "his" probability because another person may have an entirely different value for the probability of rain.)

WHICH METHOD OF ASSIGNING PROBABILITIES SHALL WE USE? The question now is: Which of the three methods are we to use? There is no simple answer, for each situation must be considered by itself. In many cases we make assumptions that allow us to determine the probability measure in a simple, straightforward manner. For example, in coin tossing it is customary to assume, unless there is evidence to the contrary, that the coin is balanced and that, using the classical method, the probabilities assigned to heads and tails are each $\frac{1}{2}$. (This sort of assumption is made in most of the games of chance we use as illustrations in this book.) The assumption of a balanced coin might not be precisely confirmed by the result of a long series of tosses. However, in the majority of cases the probability thus found would not differ greatly from that obtained by the classical method on the assumption of a balanced coin. Subjective probability results would also often agree. If the coin being tossed is known to be seriously out of balance, then an empirical determination would have to be made to obtain a suitable probability measure.

Suppose all three methods are applied and give three different values for a certain probability. Which is "correct"? Really, none is. Probabilities must be determined by whichever method appears to be appropriate to the practical problem being considered. Most often this would be the classical method or the empirical method. Less often subjective methods might be used. The assignment of probabilities is, however, a philosophical question that can never really be resolved. In spite of this we can still make progress. In the study of probability theory we shall be mainly interested in probabilistic behavior under certain strict assumptions. Certain basic probabilities will be assumed, but we shall not be concerned with how these basic probabilities were actually determined, only with how they fit into the general theory.

Note. The terms *probability* and *possibility* are sometimes confused. *Probability* refers to the likelihood of occurrence of an outcome of a phenomenon and has various levels. Thus we could say that the probability of an elephant running down Main Street, Townsville, U.S.A., is small (or large), meaning that such an outcome is unlikely (or likely) to occur. *Possibility*, on the other hand, refers to whether an outcome actually can or cannot occur. Thus we say that the possibility of an elephant running down Main Street exists (or does not exist), meaning that such a thing can (or cannot) actually occur. (If it *can* occur, we can *then* ask with what probability it does occur.) We say that a random phenomenon has a set of *possible* outcomes, and that one outcome is more *probable* than another. The terms probable and improbable are synonyms, respectively, for likely and unlikely, whereas the terms possible and impossible are synonyms, respectively, for can occur and cannot occur.

Exercises

1. A single fair die is thrown. Use the classical method to obtain the probabilities of throwing
   - (a) a six.
   - (b) an even number.
   - (c) a number less than 3.
   - (d) a number greater than 2.
2. Suppose a single card is drawn from a bridge deck. Give the probability that it will be
   - (a) a heart.
   - (b) a king.
   - (c) a face card.
   - (d) a card below 6.
3. A box contains 20 red, 30 white, and 50 blue balls; one ball is drawn from the box. State the probability that the ball is
   - (a) red.
   - (b) white.
   - (c) blue.

4. There are 50 states in the union; one is selected at random. State the probability that the selected state
    (a) has a name beginning with W.   (d) lies north of 50°N latitude.
    (b) has only four letters in its name. (e) has a rectangular shape.
    (c) borders on the Pacific Ocean.
5. Toss a coin 100 times and determine the relative frequency of heads. Then toss the same coin an additional 100 times and determine the relative frequency for the 200 tosses. Has the ratio changed greatly? On the basis of this experiment, what would you take as the probability of getting heads when tossing this coin? How does this value compare with that given by the classical method?
6. Throw a die 100 times and determine the relative frequency of each face. Repeat an additional 100 times and again determine the relative frequencies. Hence give your "estimate" of the probabilities of throwing 1, 2, 3, 4, 5, or 6. What is the sum of these probabilities?
7. A Midwestern city has 8 German restaurants, 3 French restaurants, 5 Italian restaurants, 4 Mexican restaurants, 3 Chinese restaurants, and 2 Swedish restaurants. Six people going out for dinner choose a restaurant at random from these 25. Find the probability that they will eat at a
    (a) German restaurant.   (c) "European" restaurant.
    (b) French restaurant.
8. Use the classical method to obtain the probability of
    (a) selecting a senior in a random selection of one student from a group of 10 freshmen, 15 sophomores, 20 juniors, and 25 seniors.
    (b) obtaining a sum of exactly 6 in one toss of two dice.
    (c) drawing a face card (king, queen, or jack) when one card is drawn from an ordinary deck of 52 cards.
    (d) getting exactly one head in two tosses of a balanced coin.

1.5. Methods of Enumeration

In many probability problems, we shall want to count the number of different outcomes a phenomenon can have. If we are studying the throw of a single die, it is easy to count and list the six possible outcomes: 1, 2, 3, 4, 5, and 6. However, suppose we obtain a bridge hand by selecting 13 cards at random from a deck of 52 cards. How many different bridge hands are possible? The number is so large that it would be impossible for one person to write down all the different hands and count them one by one. What we need is a more sophisticated way of counting. The following useful methods of enumeration will help us to make counts of this type very easily.

A BASIC PRINCIPLE. First we state a basic principle: If procedure 1 can be performed in n_1 ways and if for each of these ways, procedure 2 can be performed in n_2 ways, then the compound procedure "procedure 1 followed by procedure 2" can be performed in n_1n_2 ways. For example, if a coed has three green skirts and four green sweaters, she can wear 12 different skirt–sweater combinations for St. Patrick's Day.

This principle can be extended to any number of procedures. For example, if one can go from Milwaukee to Chicago by two routes ($n_1 = 2$), from Chicago to St. Louis by three routes ($n_2 = 3$), and from St. Louis to Kansas City by two routes ($n_3 = 2$), then one can go from Milwaukee to Kansas City via Chicago and St. Louis by $n_1n_2n_3 = 2 \times 3 \times 2 = 12$ different routes.

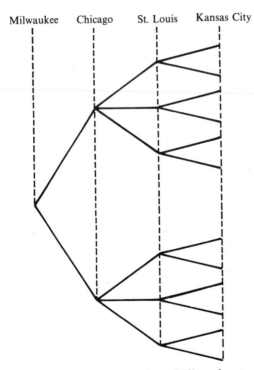

Milwaukee   Chicago   St. Louis   Kansas City

Figure 1.3. Tree diagram of routes from Milwaukee to Kansas City.

A tree diagram may be used to illustrate the principle. For the cities example, the tree diagram is given in Figure 1.3. The 12 different routes are the 12 branches beginning at Milwaukee and ending at Kansas City.

FACTORIALS. A convenient and time-saving piece of notation is the factorial symbol, which is an exclamation point, !. When we write a positive integer followed by an exclamation point, it is shorthand for "multiply together all the integers 1, 2, 3, . . . , up to the integer before the exclamation point." So, for example,

$$3! = 1 \times 2 \times 3 = 6 \qquad n! = 1 \times 2 \times 3 \times \cdots \times n.$$

Factorials grow large very rapidly, as we can see from the following calculations:

$0! =$	1	$7! =$	5,040
$1! =$	1	$8! =$	40,320
$2! =$	2	$9! =$	362,880
$3! =$	6	$10! =$	3,628,800
$4! =$	24	$11! =$	39,916,800
$5! =$	120	$12! =$	479,001,600
$6! =$	720	$13! =$	6,227,020,800.

Note that $0!$ is defined as 1, for completeness. This makes it consistent with the property $n \Pi (n - 1)! = n!$, or $(n - 1) = n!/n$. Thus when $n = 1$, $0! = 1!/1 = 1$. [The factorial notation is not used for negative numbers and so, for example, the expression $(-3)!$ has no meaning for us. The so-called gamma function enables us to extend the idea of factorials to quantities other than positive integers, but this is beyond the scope of this book.]

The factorial notation enables us to count outcomes very efficiently as we shall now see.

PERMUTATIONS. Suppose we have n different objects to arrange in n positions in a row. In how many ways can we do it? The first position can be filled by any one of the n objects, the second position by any one of the $(n - 1)$ remaining objects, and so on. By the basic principle, then, we can accomplish the arrangement in $n(n - 1)(n - 2) \cdots 1 = n!$ ways. If we use the notation $(n)_n$ to denote the number of permutations of n objects taken n at a time, we can then write $(n)_n = n!$

Suppose, instead, there are only $r \leq n$ positions but there are still n objects. How many different arrangements are possible now? A similar argument leads to the result

$$(n)_r = n(n-1)(n-2)\cdots(n-r+1) = \frac{n!}{(n-r)!},$$

the last step being easy to verify. Note that if we set $r = n$, we recover the previous result, as we should expect. We talk about "the number of permutations of n objects taken r at a time."

It is convenient to express the different arrangements of n objects in r positions as ordered r-tuples. The reader is probably already familiar with the use of ordered pairs (x, y) to represent points in a plane. If $n = 2, r = 2$, the arrangement of two objects a and b in the two positions can be expressed as an ordered pair. Thus the ordered pair (a, b) means that the object a is in the first position and the object b is in the second position. The pair (b, a) indicates that b is now in the first position and a is in the second position, so this arrangement is different from the original arrangement. We have now enumerated all the $(2)_2 = 2$ possible arrangements for this case. One of the possible arrangements of the four $(n = 4)$ letters a, b, c, and d in four $(r = 4)$ positions is given by the ordered quadruple (d, a, c, b); there are, altogether, $(4)_4 = 4! = 24$ possible arrangements of the four letters in four positions. If, however, we are arranging the four $(n = 4)$ letters among three $(r = 3)$ positions, then we represent the arrangements as ordered triples. One such triple is (b, d, a); another is (a, c, b). There are $4!/1! = 24$ distinct arrangements or triples altogether.

COMBINATIONS. We again suppose we have n objects. Now, however, instead of arranging r of the objects in a row of length r where the exact position is important, we merely want to choose r out of the n; the order in which they are arranged is irrelevant. How many such arrangements are there? In other words, what is the number of combinations of n objects taken r at a time? We first proceed as above and choose r of the n objects for the r positions. This can be done, as we have seen, in $(n)_r = n!/(n-r)!$ ways. However, the order is not important, so all rearrangements of the same r objects can be treated as identical for our purposes. Since there are $(r)_r = r!$ of these arrangements we have counted each group of r objects a total of $r!$ times instead of once in working out $(n)_r$. We must, therefore, divide this figure by $r!$ The required number of groupings, therefore, is

$$\frac{(n)_r}{r!} = \frac{n!}{(n-r)!r!}.$$

If we use the notation $\binom{n}{r}$ to denote the number of combinations of n objects taken r at a time we can then write

$$\binom{n}{r} = \frac{n!}{(n-r)!r!}.$$

Example. As an illustration, consider the $n = 5$ letters a, b, c, d, and e. How many different permutations of $r = 3$ letters can we select from these? The answer is $(n)_r = (5)_3 = 5!/2! = 60$. These 60 permutations are as follows:

abc	bca	cab	acb	cba	bac
abd	bda	dab	adb	dba	bad
abe	bea	eab	aeb	eba	bae
acd	cda	dac	adc	dca	cad
ace	cea	eac	aec	eca	cae
ade	dea	ead	aed	eda	dae
bcd	cdb	dbc	bdc	dcb	cbd
bce	ceb	ebc	bec	ecb	cbe
bde	deb	ebd	bed	edb	dbe
cde	dec	ecd	ced	edc	dce.

How many different combinations of $r = 3$ letters can we select from the $n = 5$ letters? The answer is $\binom{n}{r} = n!/[(n-r)!r!] = 5!/(2!3!) = 10$. We can get these 10 combinations by taking just one grouping of three letters out of each row of the list of permutations above. For we can see that, in each row above, each group of three letters contains the *same three letters*, differently ordered. If the order is not important, then all the $r! = 3! = 6$ groupings on a line are exactly equivalent. (If the order *is* important, of course the six groupings on a line must be regarded as different.)

This example illustrates the fact that $\binom{5}{3} = \frac{(5)_3}{3!}$.

A THEOREM ABOUT COMBINATIONS. The following result is true:

$$\binom{n}{0} + \binom{n}{1} + \binom{n}{2} + \cdots + \binom{n}{r} + \cdots + \binom{n}{n} = 2^n.$$

In words, the total number of ways of selecting $0, 1, 2, \ldots, r, \ldots,$ or n, objects from n objects is 2^n.

Proof. Consider the first object. This provides us with two possibilities when we want to select a group of objects; we can either leave it out or include it in the group. Similarly, each other object also provides two possibilities, so that altogether we have, by the basic principle, $2 \times 2 \times \cdots \times 2 = 2^n$ total possibilities in forming a group of objects. But the left-hand side above is the total number of ways we can form groups of size $0, 1, 2, \ldots, n$. No other possibilities exist. It follows that the left-hand side must equal 2^n. We can also prove this by appealing to the binomial theorem, as we shall see later (page 89).

Example. Let $n = 4$. Then

$$\binom{4}{0} + \binom{4}{1} + \binom{4}{2} + \binom{4}{3} + \binom{4}{4} = 1 + 4 + 6 + 4 + 1 = 16 = 2^4.$$

Table 1.2. $\binom{n}{r}$ **Evaluated for Certain Values of n and r**

n	2	3	4	5	6	7	8	9	10
4	6								
5	10								
6	15	20							
7	21	35							
8	28	56	70						
9	36	84	126						
10	45	120	210	252					
11	55	165	330	462					
12	66	220	495	792	924				
13	78	286	715	1,287	1,716				
14	91	364	1,001	2,002	3,003	3,432			
15	105	455	1,365	3,003	5,005	6,435			
16	120	560	1,820	4,368	8,008	11,440	12,870		
17	136	680	2,380	6,188	12,376	19,448	24,310		
18	153	816	3,060	8,568	18,564	31,824	43,758	48,620	
19	171	969	3,876	11,628	27,132	50,388	75,582	92,378	
20	190	1140	4,845	15,504	38,760	77,520	125,970	167,960	184,756

AN IMPORTANT PROPERTY OF COMBINATIONS. The following fact is often useful:

$$\binom{n}{r} = \frac{n!}{(n-r)!\,r!} = \binom{n}{n-r}.$$

Thus, for example,

$$\binom{10}{8} = \binom{10}{2} = \frac{10 \times 9}{2 \times 1} = 45.$$

In Table 1.2 we give some combinations $\binom{n}{r}$ for certain values of n and r. Since $\binom{n}{1} = n$ we omit this column. Entries that can be evaluated by the result above are also omitted; for example, $\binom{5}{3}$ is omitted because it equals $\binom{5}{2}$, which is given. For practice the reader should check some of the table entries.

PERMUTATIONS WHEN NOT ALL OBJECTS ARE DISTINGUISHABLE. Suppose we have n objects, n_1 of which are of one kind and are indistinguishable, n_2 of a second kind, $\dots, n_k$ of a kth kind, where $n_1 + n_2 + \cdots + n_k = n$. The number of different permutations of these n objects is then

$$n!/(n_1!\,n_2! \cdots n_k!).$$

This can be seen as follows. The n objects can be arranged in $n!$ ways. However, all arrangements in which the n_1 objects of the first kind are in the same set of n_1 positions can be treated as identical, and since there are $n_1!$ of these arrangements, we must divide by $n_1!$. The same reasoning applies to the n_2 objects of the second kind, and so on to the n_k objects of the kth kind. Hence the result.

Example 1. Suppose we have six objects a, a, a, b, b, c. Here we can take $n_1 = 3, n_2 = 2, n_3 = 1; n_1 + n_2 + n_3 = 6$. Then we have 60 possible different permutations given by the formula

$$\frac{n!}{n_1!\,n_2!\,n_3!} = \frac{6!}{3!\,2!\,1!} = 60.$$

To help understand why we should divide by (for example) 3!, we can label the a's as a_1, a_2, and a_3. Consider the $3! = 6$ arrangements

$$a_1 \quad b \quad c \quad a_2 \quad b \quad a_3$$
$$a_2 \quad b \quad c \quad a_3 \quad b \quad a_1$$
$$a_3 \quad b \quad c \quad a_1 \quad b \quad a_2$$
$$a_1 \quad b \quad c \quad a_3 \quad b \quad a_2$$
$$a_3 \quad b \quad c \quad a_2 \quad b \quad a_1$$
$$a_2 \quad b \quad c \quad a_1 \quad b \quad a_3.$$

These are equivalent if $a_1 = a_2 = a_3 = a$. Thus we have counted the (a, b, c, a, b, a) arrangement $3! = 6$ times over when we write down the numerator $6!$. The same is true for every other possible arrangement of the a's. Thus we must divide the $6!$ by $3!$. A similar argument on the b's and on c gives rise to the $2!$ and $1!$ in the denominator.

Example 2. Suppose we want to arrange five apples, three oranges, and four peaches in a row on a display shelf. The number of different arrangements, where all apples are alike, all oranges are alike, and all peaches are alike, is

$$12!/(5!3!4!) = 27,720.$$

Example 3. (a) Consider the problem of choosing 5 students from a class of 25 to work five different problems at the board. In how many ways can this be done? The assignment of a student to a problem makes the order important and so we want the number of permutations:

$$(25)_5 = 25!/20! = 25 \times 24 \times 23 \times 22 \times 21 = 6,275,600.$$

(b) Now suppose we want all five students to work the same problem. Now that order is not involved we want the number of combinations,

$$\binom{25}{5} = 25!/(5!20!) = 53,130,$$

a considerably smaller number.

(c) Another point to note is the following. For problem (a) we could (1) choose a set of five students as in (b) and then (2) assign the problems. The two stages are equivalent to the original procedure in (a), for we can choose the five students in $\binom{25}{5}$ ways and we can then assign the five problems to the 5 students in $5!$ ways. By the basic principle we can accomplish both stages in

$$\binom{25}{5} \times 5! = \frac{25!}{5!20!} \times 5! = (25)_5$$

ways.

Note. Many of the examples and exercises in this book involve the selection of elements from a whole collection of elements. This can be done either "with replacement," by which is meant that the same element may be included more than once in a selection, or "without replacement." For example, if two cards are to be selected from a deck, the two cards may be selected at the same time, or first one card is selected and then, without replacing the first card, the second card is selected. Both procedures are selection without replacement, for it is not possible to select the same card twice. If the first card is replaced in the deck before the second card is selected, so that the card selected first is a candidate for selection again, the procedure is selection with replacement.

Enumeration by combinations and permutations is especially useful in the case of selection without replacement. In the case of selection with replacement, the basic principle on page 17 usually suffices.

Example 1.   Three cards are selected without replacement from a bridge deck. How many "three-card hands" are possible?

If order is not important, then, since the selection is made without replacement, the number of possible hands is $\binom{52}{3} = 22,100$. If order is important, the number of possible hands is $(52)_3 = 132,600$. The latter result can be obtained, also, by using the basic principle, for there are 52 ways to select the first card, 51 ways to select the second card, and 50 ways to select the third card, giving $52 \cdot 51 \cdot 50 = 132,600$ possible hands.

Example 2.   Three cards are selected with replacement from a bridge deck. How many "three-card hands" are possible?

Since the selection is made with replacement, we apply the basic principle. There are 52 ways to select the first card, 52 ways to select the second card, and 52 ways to select the third card, giving $(52)^3 = 140,608$ possible hands, if order is important. If order is not important, the number of hands is much smaller and the problem is more difficult. Perhaps the following straightforward approach is easiest to see. There will be hands of three types: (1) three cards alike, (2) two cards alike and one different, and (3) all three cards different. There will be 52 hands of the first type; $\binom{52}{2} \cdot 2 = 52 \cdot 51 = 2652$ hands of the second type, for there are $\binom{52}{2}$ ways of selecting two different cards and two ways to assign the one to be repeated; and $\binom{52}{3} = 22,100$ hands of the third type, giving a total of 24,804 possible hands.

Exercises

1. For elective courses in a certain semester a student may choose one of two psychology courses (102, 114), one of three sociology courses (103,

174, 182), and one of two history courses (149, 163). Make a tree diagram showing all the possible combinations of the three electives.

2. There are ten automobiles in a car pool. Three automobiles are needed to take a delegation to a meeting. In how many ways can the three automobiles be selected?

3. How many different numbers of five digits each are there? (Do not use zero as the first digit.)

4. A person is holding 12 playing cards in his hand. If he plays them by selecting one at a time, at random, in how many ways can he play the first 3 cards?

5. A student finds nine books he needs at the library, but he can take out only three at a time. From how many sets of three books can he choose?

6. There are 10 candidates for an office and 6 are to be selected in a primary election. How many different groups of 6 can be chosen?

7. There are six candidates from which to choose one to run for president and one for vice president. How many different slates of candidates can be formed?

8. How many different 11-letter "words" (that is, groupings) can be formed from the letters in "engineering"?

9. There are six candidates for three offices in an organization. The voting is such that the person receiving the most votes will be president, the one receiving the second highest number of votes will be secretary, and the one receiving the third highest number of votes will be treasurer. In how many ways can the offices be filled?

10. (a) How many three-digit numbers are possible using the digits 0 to 9 inclusive if each digit may be used only once in each number and the first digit may not be zero?

    (b) How many three-digit numbers are possible using the digits 0 to 9 inclusive if each digit may be used *more than once* in each number and the first digit may not be zero?

11. In how many ways can five books be arranged on a shelf?

12. How many integers can we form with three different digits, using one, two, or three digits at a time in the following cases?

    (a) Each digit may be used only once in each number.

    (b) Each digit may be used *more than once* in each number.

13. How many three-letter "words" (that is, groupings) are possible using the English alphabet in the following cases?

    (a) Each letter may be used only once in each word.

    (b) Each letter may be used more than once in each word.

14. How many groups of size 3 can be formed from 100 objects?

15. How many different committees of four can be chosen from eight people?

16. If you have five different kinds of fruit in the refrigerator and want to choose two kinds for lunch, how many combinations are possible?

17. How many different 10-letter "words" (that is, groupings) can be formed from the letters in Oconomowoc?

18. How many 6-flag signals can be formed with 3 red, 2 blue, and 1 white flag?

19. How many different license plates can be made if
    (a) the plates each have one letter followed by six digits?
    (b) the letters I, O, and Z are not used, and the first digit may not be zero?
    (c) a letter can be followed by either 1, 2, 3, 4, 5, or 6 digits, with the restrictions on letters and numbers given in (b)?

20. (a) How many lines can be drawn between seven points, no three of which lie on a straight line?
    (b) How many triangles can be formed with the seven points in (a) as vertices?

21. How many different poker hands (5 cards each) can contain
    (a) 4 aces?
    (b) 5 cards of the same suit?
    (c) 3 cards of one value and 2 of another value?

22. Ten consecutive intersections have signal lights. If, at a given instant, 5 are red, 3 are green, and 2 are yellow, how many different arrangements are possible?

23. A contest crossword puzzle has 14 word blanks for each of which there are two choices, 5 blanks for each of which there are three choices, and 1 blank for which there are four choices. How many different solutions are possible?

24. A combination door lock has five buttons that must be pressed in a particular order to open the door. The lock manufacturer advertises that there are an infinite number of combinations possible. Exactly how many are there?

25. A combination bicycle lock has numbers from 1 to 30 inclusive. The lock is opened by turning clockwise to a certain number, then counter-clockwise to a certain number, and then clockwise again to a certain number. How many combinations are possible?

26. A coin collector wants to purchase one each of the following coins: cent, nickel, dime, quarter, and half-dollar. The dealer has 6 cents, 8 nickels, 5 dimes, 4 quarters, and 3 half-dollars that the collector is interested in. How many different combinations of coins can the collector purchase?

27. A manufacturer makes prefabricated houses of one basic design. Variety is achieved by using one of three colors of siding, using one of three types of shingles, and by choosing either a garage or a carport. How many different houses can be built?

28. A dress designer has come up with a basic pattern that can be varied

by using one of three types of collar, one of four types of sleeves, and by having or not having pockets. If the dress can be made of any one of four colors of fabric, how many different dresses are possible?

29. In how many possible ways can four prizes be distributed among 12 contestants if

(a) no person may win more than one prize?

(b) there is no restriction on the number of prizes one person may win? State clearly any additional assumptions you make.

30. A freshman at a certain college must take a science course, a history course, a language course, an English course, and two theology courses. In how many ways can he choose a program if there are five science, three history, six language, two English, and seven theology courses offered?

31. A baseball manager has four pitchers and two players for each of the other eight positions. How many different teams could be put on the field?

32. From four married couples, in how many different ways can a pair of bridge partners consisting of a man and woman be chosen if no husband and wife are to play together?

33. In how many ways can four men and four women be seated in a row if men and women must occupy alternate seats? How many if there are no restrictions except that a man must sit in the first seat?

34. In how many different ways can eight different jobs be filled by 12 applicants if any applicant may be assigned to any job?

35. In how many ways may a team of eight workers be formed from a group of 12 applicants?

36. How many different 13-card bridge hands may be dealt from a deck of 52 cards? How many of these hands contain four aces?

37. How many different 5-card hands can be dealt from a bridge deck?

38. In how many ways can 15 distinguishable objects be separated into three piles of equal numbers of objects?

39. In how many ways can six married couples be seated around a table so that men and women alternate?

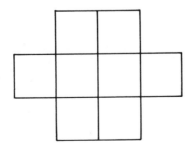

40. The eight basic squares in the figure on p. 27 are to be filled with the integers 1 to 8 inclusive, one to a square and all different. How many different arrangements are possible? If no two consecutive numbers are to be adjacent horizontally, vertically, or diagonally, how many different arrangements are possible now?

CHAPTER 2
Sets and Subsets

2.1. Definitions and Notation

The concept of a set is basic in many branches of mathematics, and the terminology and notation of sets are particularly useful and convenient for introducing the mathematical theory of probability. We now give a brief introduction to those aspects of set theory we shall use in this book.

Definition. A set is a well-defined collection of distinct objects (usually called the elements of the set).

This means that the designation of a collection of elements as a set must be done in such a way that, if we are given any object, it must be possible to ascertain whether or not the object belongs to the set. In other words, a set must be described in an unambiguous manner.

The set of letters a, c, i, s, t can be described as the set of distinct letters of the word "statistics." Note that although statistics contains the letter s three times, the letter t three times, and the letter i twice, the set lists each letter only once. (Of course, the set of letters in the word "attics" or in "static" is exactly the same set.) Now, if we are given any letter of the alphabet—or any other object—we can state either that the letter or object belongs to the set or that it does not belong to the set.

NOTATION. It is convenient to represent a set by a capital letter (for example, A) and its elements by lowercase letters (for example, $a, b, c, \ldots$). We can use braces ("curly" brackets) to enclose the elements of the set (or a description of the elements). The set of distinct letters in "statistics" above might be designated A and we could write $A = \{a, c, i, s, t\}$, where the equals sign merely indicates that A represents the set of letters described. This also serves as a sufficient description of the set. "The set of letters in the word statistics" and "the set $\{a, c, i, s, t\}$" are equivalent.

If the element a belongs to the set A, we write $a \in A$ and we read: "small a belongs to big A" or "small a is an element of the set big A."

Additional examples of sets are:
1. The players in the National Football League.
2. The days of the week.
3. The real numbers.
4. The college freshmen in the United States.
5. The chemical elements listed on the periodic chart.
6. The publications in the Library of Congress.
7. The voters in the last U.S. presidential election.

Sets may be finite or infinite. A finite set is one containing a finite number of elements; all other sets are infinite. Of the sets listed above, only set 3 is infinite. The set W of all positive integers is also infinite. Since we cannot actually list all the elements of W, an alternative way of describing it is: $W = \{x : x \text{ is a positive integer}\}$. The colon symbol ($:$) is read "such that" or "where." This method of describing a set by using a variable and a defining property can be used for any set, infinite or finite. In the "statistics" example above, we could write: $A = \{x : x \text{ is a letter of the word "statistics"}\}$. (It was easier to write simply: $A = \{a, c, i, s, t\}$, of course, but for some sets the other description would be preferable.)

Exercises

1. Write (in a form using braces)
    (a) the set of letters in the word "inconvenient."
    (b) the set of real numbers between 1 and 5 inclusive.
    (c) the set of presidents of the United States from Wilson to Truman inclusive.
    (d) the set of planets in our solar system.
    (e) the set of all integers.
2. Tell which of the sets in Exercise 1 are finite and which are infinite.
3. Give two examples of finite sets and two examples of infinite sets that are different from those in the text and those in Exercise 1.

2.2. Subsets

Any number of the elements of a set A is said to form a subset of A. That is, set B is a subset of A if every element of B is an element of A. This can be expressed as follows: B is a subset of A if, for every $x \in B$, we have $x \in A$. Subsets are also indicated by capital letters and their elements are enclosed in braces because they, too, are sets. For example, the players with the Green Bay Packers form a subset of the players in the National Football League. So do all the centers in the NFL, or all players whose last names begin with S, or all players who are married, or all those who have 20-20 vision. Note that the set of Green Bay centers forms a subset of the set of all Green Bay players; the former set is thus a subset of a subset!

Subsets of the set $A = \{a, b, c\}$ are $\{a\}$, $\{b\}$, $\{c\}$, $\{a, b\}$, $\{a, c\}$, $\{b, c\}$, $\{a, b, c\}$, and $\{\ \}$, the empty, or null, set. The null set, being a set with no elements, is a subset of every set. It is given the special symbol $\varnothing$; thus $\varnothing = \{\ \}$.

PROPER SUBSETS. All subsets of a set A except the set A itself are called proper subsets of A. A set is an *improper* subset of itself.

Example. In the previous example, $\{a, b, c\}$ is an improper subset of A; the other subsets given are proper subsets.

To indicate that B is a subset (proper or improper) of A we can write $B \subseteq A$ or, equivalently, $A \supseteq B$. When it is necessary to distinguish between proper subset and just subset, we shall write $B \subset A$ when B is a proper subset of A and $B \subseteq A$ when B could be improper.

Examples. Let $A = \{a, b, c, d\}$, $B = \{a, b, c\}$, $C = \{a, c\}$. Then since B is a proper subset of A we write $B \subset A$, or $A \supset B$. Since C is a proper subset of both A and B we write $C \subset A$ and $C \subset B$. Since every set is a subset of itself we can write $A \subseteq A$, $B \subseteq B$, and $C \subseteq C$. Since the null set, $\varnothing$, is a proper subset of every set, we can write $\varnothing \subset A$, $\varnothing \subset B$, and $\varnothing \subset C$.

We shall now drop use of the words proper and improper because they are contained in our notation.

Exercises

1. (a) Write a subset of size 3 of the set of letters in the word "inconvenient"; how many such subsets are there? How many are there of size 6?
   (b) Write an infinite subset of the set of real numbers between 1 and 5 inclusive; now write a finite subset of size 400.
   (c) Write a nonempty proper subset of the set of presidents of the United States from Wilson to Truman inclusive.
   (d) Is there a subset common to all three of the sets in (a), (b), and (c)?
2. Give a subset of each of the sets listed on page 30.

2.3. Set Properties and Operations

SIZE OF A SET. It was stated earlier that sets may be finite or infinite. An important characteristic of a finite set A in probability theory is its size, which we designate by the symbol $n(A)$, and which denotes the number of elements in the set. So, for example, if $A = \{a, b, c\}$, then $n(A) = 3$. Obviously, $n(\varnothing) = 0$.

ONE-TO-ONE CORRESPONDENCE. If $n(A) = n(B)$, sets A and B can be placed in one-to-one correspondence. This means that, for each element in A, we can specify a corresponding element in B, and vice versa. For example, the sets $A = \{h, i, r\}$ and $B = \{19, 3, 7\}$ can be placed in one-to-one correspondence. The correspondence may be $h \leftrightarrow 19$, $i \leftrightarrow 7$, $r \leftrightarrow 3$, where the arrows indicate the partners in the correspondence, or any of the other five possible associations. Infinite sets can be placed in one-to-one correspondence in a similar way.

EQUALITY OF SETS. If every element of A is identical to an element of B, and vice versa, then we write $A = B$. That is, $A = B$ if and only if $A \subseteq B$ and $B \subseteq A$. As an illustration, consider the set of letters in "statistics" and the set of letters in "attics." These sets are equal, because each contains the same five letters $\{a, c, i, s, t\}$.

THE UNIVERSAL SET. In order to define and discuss operations among sets, it is necessary first to define an overall reference set, called the *universal set* and denoted by U.

U is the set of all elements under consideration in a particular discussion, and all sets we wish to talk about in a certain exercise are subsets of U. For example, if we wished to talk about certain subsets of real numbers, the set of all real numbers would serve as the universal set U. If we are considering subsets of integers, then the set of all integers would serve as the universal set.

Now that the universal set has been defined, we may define certain set operations. The basic operations on sets are forming the complement, the union, and the intersection.

COMPLEMENT. The complement A' of a set A relative to the universal set U is the set of all elements in U but not in A. Thus, if $x \in A$, then $x \notin A'$ (read x does not belong to A') and if $y \in A'$, then $y \notin A$.

Example 1. The complement of $A = \{a, b\}$ relative to the set $U = \{a, b, c\}$ is the set $A' = \{c\}$.

Example 2. Let $H =$ heads and $T =$ tails. The complement of $A = \{H\}$ when tossing an ordinary coin is $A' = \{T\}$, the universal set being $U = \{H, T\}$.

Example 3. Suppose we toss a die. The numbers (or "points") 1 to 6 form the set U of all possible outcomes. Suppose $A = \{2, 4, 6\} = \{x : x \text{ is even}\}$. Then the complement A' of A is the set $A' = \{1, 3, 5\} = \{x : x \text{ is odd}\}$.

Example 4. The complement of the set of rational numbers relative to the set of real numbers is the set of irrational numbers.

VENN DIAGRAM. A convenient device for illustrating sets graphically is the *Venn diagram*. Often (but not necessarily) the universal set U is represented by a rectangle, and other sets, subsets of U, by circles or by other enclosed regions of the plane. In the Venn diagram of Figure 2.1 we see a given set A, represented by a circle, lying within the rectangle representing U. The complement of A, namely A', is the region of U outside the circle and within U.

UNION. The union of two sets A and B, written $A \cup B$, is the set of all elements belonging either to A or to B. (Some of these may belong to both A and B.) Thus, if $x \in A$, then $x \in A \cup B$, and if $y \in B$, then $y \in A \cup B$. The union of $A = \{a, b, c\}$ and $B = \{b, c, d\}$ is $A \cup B = \{a, b, c, d\}$. The union of the set of rational numbers and the set of irrational numbers is the set of real numbers, for all real numbers are either rational or irrational and hence belong to one or the other of the sets forming the union. The union of the set of rational numbers and the set of integers is the set of rational numbers.

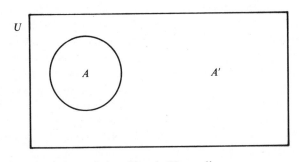

Figure 2.1. Simple Venn diagram.

The union of A and its complement A' is the universal set U; thus $A \cup A' = U$. Furthermore, $A \cup U = U$ and $A \cup \emptyset = A$, as can be verified. (*Note:* Since the *U*nion symbol is shaped like a U it is easy to remember.)

INTERSECTION. The intersection of two sets A and B written $A \cap B$, is the set of elements belonging to both A and B. Thus $x \in A \cap B$ if $x \in A$ and $x \in B$. The intersection of $A = \{a, b, c\}$ and $B = \{b, c, d\}$ is $A \cap B = \{b, c\}$. The intersection of the set of rational numbers and the set of integers is the set of integers, for integers are rational numbers as well as integers. The intersection of the set of rational numbers and the set of irrational numbers is the null set, $\emptyset$, for there is no number that is both rational and irrational.

DERIVATION OF OTHER SETS. It is obvious that the operations of forming the complement, the union, and the intersection produce sets, and of course these same operations can be performed again on the new sets. For example, combinations such as $A \cup (B \cap C)$, $A' \cap B'$, and $(B \cup C)'$ are perfectly meaningful and are, once again, sets themselves.

All these derived sets are necessarily subsets of the universal set. For example, suppose A and B are subsets of the universal set. Is the set $(A' \cap B') \cup (A \cap B)$, for example, a subset of U? A review of the definitions will show that if, A and B are any two subsets of U, then A' and B' are subsets of U, so both $A' \cap B'$ and $A \cap B$ are also subsets of U. Hence, finally, the union of the last two subsets is also a subset of U. More formally we can write the following. If $x \in (A' \cap B') \cup (A \cap B)$, then $x \in A' \cap B'$ or $x \in A \cap B$. If $x \in A \cap B$, then $x \in A$ and $x \in B$, and so $x \in U$, since $A \subset U$ and $B \subset U$. If $x \in A' \cap B'$, then $x \in A'$ and $x \in B'$, so that $x \in U$ by definition of the complement set.

A POINT OF NOTATION. We now pause briefly to make an important, if rather pedantic point, about our notation. The two statements $x \in A$ and $\{x\} \subset A$ look similar but do not mean the same thing. The first, $x \in A$, means that the *single element* x belongs to the set A. The second, $\{x\} \subset A$, means that the *set consisting of the single element x is a subset* of the set A. In other words, in the first notation x is an element and in the second $\{x\}$ is the set containing the element x. To say it yet another way: In the first case we are saying that the element x satisfies the definition of set A and therefore belongs to the set A. In the second we are saying that set A has been defined and therefore has subsets, and that $\{x\}$ is one of these subsets. The element x can have no subsets, no complement relative to U, and no union or intersection with sets; the set $\{x\}$, on the other hand, has subsets $\{\ \}$ and $\{x\}$, has a complement relative to U, and its union and intersection can be taken with other subsets of U. (Note that the symbol $\subset$ is used *only between two sets*.)

DISJOINT SETS. If it happens that $n(A \cup B) = n(A) + n(B)$, then the sets A and B are said to be *disjoint*, and it follows that $A \cap B = \emptyset$. Note that $\emptyset$ is the null set and not a zero element. However, when A and B are disjoint, $n(A \cap B) = n(\emptyset) = 0$, an ordinary numerical zero! Disjoint sets have only one common subset, the null set. Two sets, A and B, whether finite or infinite, are disjoint if $A \cap B = \emptyset$.

If A and B are *not* disjoint, that is, if A and B have at least one element in common (so that A and B have a common nonempty subset), then we say that A *meets* B.

For any two finite sets A and B, disjoint or not disjoint,

$$n(A \cup B) = n(A) + n(B) - n(A \cap B).$$

This is easily seen by means of the Venn diagram in Figure 2.2. The elements in $A \cap B$ are included in A and also in B, and so are counted twice, once in $n(A)$ and once in $n(B)$. Subtracting $n(A \cap B)$ corrects this double count.

Exercises

1. Given sets $A = \{4, 6, 9, 13\}$ and $B = \{2, 3, 9\}$, and $U = \{x : x$ is a positive integer less than 15$\}$, find
   (a) $A \cup B$.
   (b) $A \cap B$.
   (c) $n(A)$.
   (d) $n(B)$.
   (e) $n(A \cup B)$.
   (f) A' (complement of A).
   (g) $(A \cup B)'$.
   (h) $A' \cap B'$.
   (i) $(A \cap B)'$.
   (j) $A' \cup B'$.
2. Draw Venn diagrams illustrating each of the following sets:
   (a) $A \cup B$.
   (b) $A \cap B$.
   (c) $(A \cup B)'$.
   (d) $(A \cap B)'$.
   (e) A'.
   (f) $A' \cup B'$.
   (g) $A' \cap B'$.

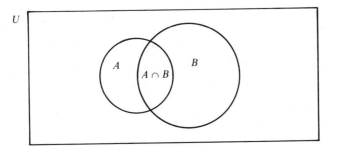

Figure 2.2. Finding the size of a union set.

2.4. Additional Properties of Sets

DIFFERENCE OF SETS. The difference $A - B$ of two sets A and B is defined to be the set of elements belonging to A but not to B. Thus $x \in A - B$ if $x \in A$ and $x \notin B$. Taking the difference of two sets A and B is not a new set operation because it is equivalent to taking the intersection of A and B'; that is, $A - B = A \cap B'$, as is easily proved.[1]

The difference of sets is a useful operation for separating a universal set into disjoint subsets. For example, if A meets B we represent the situation by a Venn diagram as in Figure 2.3. It is seen that the contours within U divide U into four disjoint subsets: $A - B$, $B - A$, $A \cap B$, and $(A \cup B)'$. These are the four basic disjoint subsets of U for any pair of subsets A and B. For additional examples of Venn diagrams see Figure 2.4.

ASSOCIATIVE AND SIMILAR PROPERTIES. Sets and set operations have properties similar to those of the real number system. For example, the operation of forming the union of sets is associative,[2] namely

$$A \cup (B \cup C) = (A \cup B) \cup C \qquad (= A \cup B \cup C) \qquad \text{(see Figure 2.5).}$$

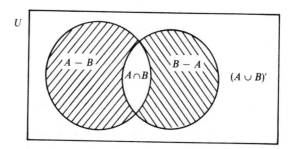

Figure 2.3. Venn diagram with four disjoint subsets.

---

[1] $x \in A - B$ means that $x \in A$ and $x \notin B$; but $x \notin B$ means that $x \in B'$, and therefore, since $x \in A$ and $x \in B'$, we have $x \in A \cap B'$. Thus every element of $A - B$ is in $A \cap B'$. The converse also follows easily.

[2] This can be seen as follows. Let x be an element of the set $A \cup (B \cup C)$; that is, $x \in A \cup (B \cup C)$. Then $x \in A$ or $x \in (B \cup C)$, and therefore $x \in A$ or $x \in B$ or $x \in C$. If $x \in A$, then $x \in (A \cup B)$ and $x \in (A \cup B) \cup C$. If $x \in B$, then $x \in (A \cup B)$ and $x \in (A \cup B) \cup C$. If $x \in C$, then $x \in (A \cup B) \cup C$. Therefore, by the definition of subset, $A \cup (B \cup C) \subseteq (A \cup B) \cup C$. Similarly, $(A \cup B) \cup C \subseteq A \cup (B \cup C)$, and the result follows, because every element of $A \cup (B \cup C)$ is an element of $(A \cup B) \cup C$ and vice versa. A Venn diagram is helpful here; see Figure 2.5.

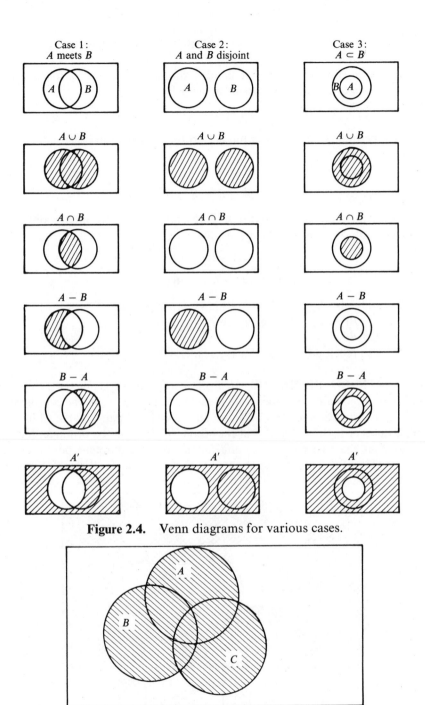

Figure 2.4. Venn diagrams for various cases.

Figure 2.5. Shaded area is $A \cup B \cup C$.

37

It can be shown in a similar manner that the operation of forming the intersection of sets also is associative, namely

$$A \cap (B \cap C) = (A \cap B) \cap C \qquad (= A \cap B \cap C).$$

Both operations, forming the union and forming the intersection, are also commutative, that is,

$$A \cup B = B \cup A,$$

$$A \cap B = B \cap A.$$

Furthermore, the distributive law holds for union over intersection and intersection over union. This means that

$$A \cup (B \cap C) = (A \cup B) \cap (A \cup C),$$

$$A \cap (B \cup C) = (A \cap B) \cup (A \cap C).$$

The following laws, de Morgan's laws, are important:
1. $(A \cup B)' = A' \cap B'$ (see Figure 2.6).
2. $(A \cap B)' = A' \cup B'$.
Proofs are left as exercises for the reader.

CARTESIAN PRODUCT.   In algebra and analytic geometry the notion of an ordered pair of numbers is of great value. For example, the coordinates of any point in a plane can be written as an ordered pair (x, y) in which the first number is the abscissa and the second number is the ordinate. Recall, also, that the two pairs (x, y) and (y, x) represent different points unless $x = y$. (Here the symbol x represents any element of the set R of real numbers;

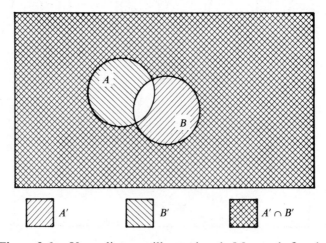

Figure 2.6.   Venn diagram illustrating de Morgan's first law.

the symbol y, likewise, represents any element of R.) In general, if we have two sets A and B we can form a set of ordered pairs, each pair taking its first element from one set and its second element from the other set. For example, if $A = \{a, b, c\}$ and $B = \{1, 2\}$, then we can form the set of ordered pairs $\{(a, 1), (a, 2), (b, 1), (b, 2), (c, 1), (c, 2)\}$. This set is called the *Cartesian product* of A and B and is written $A \times B$. Note that the product set, $A \times B$, is a set whose elements are ordered pairs of elements. Thus we define the Cartesian product, or product set, as follows:

$$A \times B = \{(x, y) : x \in A, y \in B\}.$$

In the example above, $n(A) = 3$ and $n(B) = 2$. The product set $A \times B$ has size $n(A \times B) = n(A) \cdot n(B) = 3 \cdot 2 = 6$.

Note carefully that the Cartesian product $B \times A$ is a different set consisting of elements $(1, a)$, $(1, b)$, $(1, c)$, $(2, a)$, $(2, b)$, $(2, c)$; it does, however, have the same number of elements, $n(B) \cdot n(A) = 6$.

As another example of the use of the Cartesian product consider the tossing of two dice. If for one die we write $S = \{1, 2, 3, 4, 5, 6\}$ for the set of possible outcomes, then all possible outcomes for *two* dice are represented by the set $S \times S$. Also, $n(S \times S) = n(S) \cdot n(S) = 36$.

The idea of a Cartesian product is easily extended. For example, we define the Cartesian product of three sets, $A \times B \times C$, to be the set of ordered triples (a, b, c), where a, b, and c are elements of A, B, and C, respectively. Cartesian products of more sets are defined in a parallel manner.

Exercises

1. Show that $(A - B)' = A' \cup B$.
2. Show that $(A \cup B) - (A \cap B) = (A - B) \cup (B - A)$.
3. In a group of 11 adults 5 are male, and 2 persons of each sex are single. Draw a Venn diagram in which subsets A and B represent "maleness" and "singleness" and show the numbers of adults in the four possible disjoint subsets.
4. Of 40 patrons served in a restaurant, 14 had pie; 14 had ice cream; 14 had coffee; 5 had pie and ice cream; 4 had pie and coffee; 3 had ice cream and coffee; and 1 had pie, ice cream, and coffee. Make a Venn diagram showing the number of patrons in each of the eight disjoint subsets into which the set of patrons can be divided.
5. Prove:
   (a) $(A \cap B) \cap C = A \cap (B \cap C)$.   (d) $(A \cup B)' = A' \cap B'$.
   (b) $A \cap (B \cup C) = (A \cap B) \cup (A \cap C)$.   (e) $(A \cap B)' = A' \cup B'$.
   (c) $A \cup (B \cap C) = (A \cup B) \cap (A \cup C)$.
6. Illustrate the statements in Exercise 5 by Venn diagrams.

7. Show by means of Venn diagrams (or otherwise) that
   (a) $(A \cap B') \cup (A' \cap B) \cup (A \cap B) = A \cup B$.
   (b) $(A \cap B) \cup (A \cap B') = A$.

8. Form the product set $A \times B$ for the sets $A = \{4, 6, 9, 13\}$, $B = \{2, 3, 9\}$.

9. Prove that $n(A \times B \times C) = n(A) \cdot n(B) \cdot n(C)$.

10. For sets A and B, where A meets B, show by Venn diagrams that
    (a) A and $B - (A \cap B)$ are disjoint.
    (b) $A \cup B = A \cup [B - (A \cap B)]$.
    (c) $A \cap B$ and $A - (A \cap B)$ are disjoint.
    (d) $A \cap B$ and $B - (A \cap B)$ are disjoint.
    (e) $A \cup B = (A \cap B) \cup [A - (A \cap B)] \cup [B - (A \cap B)]$.
    (f) $A \cap B' = A - B$.
    (g) $A \cap B = U - (A' \cup B')$.
    (h) $(A \cap B) \subset A$.
    (i) $(A \cap B) \subset B$.
    (j) $A \subset (A \cup B)$.
    (k) $B \subset (A \cup B)$.

11. For three sets A, B, and C that all meet each other, show that A, $B - A$, and $C - (A \cup B)$ are disjoint sets whose union is $A \cup B \cup C$.

12. S is the set of students at a certain university, E_1 is the subset of students who belong to the Young Democrats, E_2 is the subset of students who belong to the $\pi\mu\varepsilon$ fraternity, E_3 is the subset of students who belong to the university chorus. Describe the sets
    (a) $E_1 \cup E_2 \cup E_3$.
    (b) $E_1 \cap E_2 \cap E_3$.
    (c) $S - (E_1 \cup E_2)$.
    (d) $E_1' \cup E_2' \cup E_3'$.
    (e) $E_1 \cap E_2 \cap E_3'$.
    (f) $(E_1 \cup E_2) \cap E_3$.
    (g) $E_1' \cap (E_2 \cup E_3)$.
    (h) $(E_1 \cup E_2)' \cap E_3$.
    (i) $E_1 \cup (E_2 \cap E_3)$.
    (j) $E_1' \cap E_2' \cap E_3'$.

    Express in symbols, the set of students who belong to
    (k) two or more of the three organizations.
    (l) not more than one organization.

13. S is the set of all possible bridge hands; E_1 is the set of hands with all spades, E_2 is the set of hands with all hearts, E_3 is the set of hands with all diamonds, and E_4 is the set of hands with all clubs. Describe and give the number of elements in each of the sets
    (a) $E_1 \cap E_2$.
    (b) $E_1 - (E_2 \cap E_3)$.
    (c) $(E_1 \cap E_2 \cap E_3) \cup E_4$.
    (d) $E_1 \cup (E_2 \cap E_3)$.
    (e) $(E_1 \cup E_2) - E_3'$.
    (f) $E_1' \cup E_2'$.
    (g) $E_1' \cap (E_2' \cup E_3')$.
    (h) $(E_1 \cup E_2 \cup E_3) - E_4$.
    (i) $S - [(E_1 \cup E_2) \cap (E_3 \cup E_4')]$.

CHAPTER 3

The Sample Space and the Probability Function

In this chapter, we build upon the foundations laid in Chapters 1 and 2 and begin the study of probability theory by defining the important concepts of *sample space* and *probability function*.

3.1. Sample Spaces

The idea of a random phenomenon was discussed in Chapter 1. We said that a random phenomenon could have a number of possible outcomes. In Chapter 2 we saw that the universal set U is the complete set of possible outcomes of the random phenomenon being studied. This universal set will now be used to provide us with sample spaces.

Definition. Suppose the possible outcomes of a random phenomenon are characterized in some unique manner and labeled accordingly. Then the set of all the possible labels is a *sample space* or *probability space*. An element of a sample space is called a *sample point* or a *simple event* or an *elementary event*. The letter S is often used to denote a sample space.

Example 1. A single six-sided die is tossed and the six possible outcomes are characterized by the number that appears on the throw. Then the sample space is $S = \{1, 2, 3, 4, 5, 6\}$. (Note that in this case the sample points of S form the universal set, but this is not always true,[1] as demonstrated by Examples 2 and 3.) Here S has six sample points.

Example 2. A single six-sided die is tossed and the six possible outcomes are characterized by whether the number that appears on the throw is odd or even. Now the sample space is $S = \{$odd number, even number$\}$. S here has two sample points.

Example 3. A single six-sided die is tossed and the six possible outcomes are characterized by whether the number that appears on the throw is 5 or smaller or is 6. The appropriate sample space is $S = \{5$ or smaller, $6\}$. This sample space also has two sample points.

We see that the form and size of the sample space depend on the chosen characterization and that the universal set U can provide a number of different sample spaces.[2] In many applications, however, the sample space and the universal set will be identical. Note that, whatever the sample space may be, each outcome of a random phenomenon corresponds to one and only one element of the sample space S as in Examples 1, 2, and 3. The converse is not necessarily true, however; some, or all, elements of the sample space may correspond to *more* than one outcome, as in Examples 2 and 3.

EVENTS. In probability theory we are often interested in the individual sample points of a sample space. However, we often wish to consider several sample points simultaneously; hence we make the following definition: An *event* is any subset of sample points belonging to the sample space.

---

[1] Some writers define a sample space more restrictively so that this is *always* true. We have given a more general definition, however.

[2] Although the choice of sample space is, to a certain extent, arbitrary, it is usually an advantage to choose it in such a way that the computation of probabilities is made as simple as possible. For example, if we are interested in problems involving the tossing of three coins, it is usually more convenient to start with the sample space $S = \{(H, H, H), (H, H, T), (H, T, H), (T, H, H), (T, T, H), (T, H, T), (H, T, T), (T, T, T)\}$, which has eight equiprobable ($\frac{1}{8}$) sample points (by the classical definition) rather than to use the sample space $S' = \{3H, 2H$ and $1T, 1H$ and $2T, 3T\}$ with only four sample points whose probabilities are $\frac{1}{8}, \frac{3}{8}, \frac{3}{8}$, and $\frac{1}{8}$, respectively (see page 7).

(Above we called elements of the sample space "simple events." Note that an event, as defined here, could be a *simple* event consisting of one sample point, or a *compound* event consisting of several sample points. When we specifically want to draw attention to the fact that an event is only a single sample point we can call it specifically a "simple event.")

Example.   If we throw a die and define the corresponding sample space as $S = \{1, 2, 3, 4, 5, 6\}$, then $\{1\}$, $\{2, 3\}$, $\{1, 5, 6\}$, $\{\ \}$, and $\{1, 2, 3, 4, 5, 6\}$ are all events associated with this random phenomenon. There are, of course, many others.

The total number of events associated with a random phenomenon is the total number of possible subsets of the sample space S. If $n(S) = N$, then there are 2^N subsets of S and thus 2^N events in the sample space. This is clear from the argument that each simple event either belongs or does not belong to any specific event (see page 21).

We noted above that each individual element or point of the sample space is itself an event. The null set, also, is an event. We know that unions, intersections, complements, and differences of sets are also sets. Hence such combinations of events are also events, because they are subsets of the sample space. An important event is the whole sample space itself, of course.

Example.   Suppose we toss a single die and define $S = \{1, 2, 3, 4, 5, 6\}$. Let event $E_1 = \{\text{even number}\} = \{2, 4, 6\}$ and event $E_2 = \{\text{number divisible by } 3\} = \{3, 6\}$. Then if $E_3 = \{\text{number divisible by 2 or 3}\} = \{2, 3, 4, 6\}$, E_3 can be expressed as the union of E_1 and E_2, namely $E_3 = \{2, 4, 6\} \cup \{3, 6\}$. (Note again that the word "or" in the language of events means *either* of two events or *both*. Thus in E_3, the point 6 is included because it belongs to both E_1 and E_2, while the points 2, 4, and 3 belong either to E_1 or to E_2.)

MUTUALLY EXCLUSIVE EVENTS.   Two events A and B are said to be mutually exclusive if the subsets A and B are disjoint, that is, if $A \cap B = \varnothing$. (This means, of course, that A and B have no sample points in common.)

Example.   If one card is drawn from a bridge deck, the event $A = \{\text{the card is a spade}\}$ and the event $B = \{\text{the card is a heart}\}$ are mutually exclusive.

(However, the event A above and the event $C = \{\text{the card is an ace}\}$ are *not* mutually exclusive, for the card drawn could be the ace of spades; if so, the outcome would belong to both A *and* C.)

Exercises

1. Describe sample spaces for the following experiments:
    - (a) tossing three coins.
    - (b) rolling two dice.
    - (c) choosing an integer from the integers 1 to 10 inclusive.

(d) choosing a positive integer from the set of all positive integers.

(e) choosing a real number from the set of all real numbers.

2. Give sample spaces for the tossing of
   (a) two coins. (b) four coins. (c) one coin and one die.

3. Give a sample space different from one in Exercise 2(b) for the tossing of four coins.

4. A certain course is to be given on two of the five school days (Monday, Tuesday, Wednesday, Thursday, Friday). Describe a sample space of the possible outcomes.

5. From a set of four girls and three boys, one boy and one girl are to be selected for a committee. Describe a suitable sample space.

6. Two of the colors red, white, blue, yellow, and green are to be selected as class colors. Describe a sample space for the possible combinations.

7. Two dice are thrown. Give the subset of outcomes belonging to the following events in the universal set U: the sum is
   (a) 7. (c) 7 or 11. (e) 2 or 12.
   (b) 11. (d) at most 4.

8. Two dice are to be tossed and we are interested in whether the sum showing is divisible by 3.
   (a) Describe a sample space.
   (b) Are the sample points equally likely?

9. There are three slips of paper in a box; one is marked with a 4, one with a 9, and one with a W. Two slips are drawn from the box at random at the same time. Describe a sample space for this experiment. (Order is not important.) List all the events associated with the sample space.

10. A box of N bolts has $k < N$ bolts that have no threads. Bolts are randomly selected one at a time without replacement until one with threads is found.
    (a) Describe a sample space for this experiment. How many points are there?
    (b) Describe a sample space if bolts are selected until n threaded bolts are found.

11. One card is to be drawn at random from an ordinary bridge deck. Describe suitable sample spaces if we are interested in
    (a) the particular card drawn.
    (b) the suit drawn.
    (c) the value drawn.
    (d) the color drawn.
    (e) whether the card drawn is a face card.

12. Give the number of events for each of the sample spaces in Exercise 11.

13. A coin is tossed five times. How many points are there in the largest possible sample space? How many of these have two H's and three T's? How many have three H's and two T's?
14. How many possible subsets are there in a sample space with seven sample points?
15. Write down all the events associated with the sample space $S = \{a, b, c\}$.
16. Two dice are tossed. Say whether the following pairs of events are mutually exclusive:
    (a) the sum is 7; the sum is 6.
    (b) the sum is 8; one die shows a 5.
    (c) the sum is an odd number; one die shows a 6.
    (d) the sum is an even number; the sum is divisible by 3.
    (e) the sum is greater than 9; one die shows a 2.
    (f) the sum is divisible by 3; one die shows a 4.
    (g) the sum is divisible by 3; the sum is greater than 9.

3.2. Assigning Probabilities to Events Composed of Equally Likely Simple Events

We saw in Chapter 1 that, if we have a random phenomenon, the individual outcomes can be assigned probabilities in various ways. We noted in particular the classical method and the relative frequency method.

Suppose we have a particular random phenomenon and S denotes our selected sample space. Let $n(S) = N$ denote the number of sample points in S, where we assume N to be finite. If all the simple events (or sample points) in S are equally likely to occur as an outcome, we can assign each simple event a probability of $1/N$, by the classical method. (We can make the same assignment using the relative frequency method if it can be assumed that the relative frequencies of occurrence of all of the sample points are the same.)

Suppose E is a subset of S. The event E is said to occur whenever the outcome is a simple event belonging to E. If E consists of $n(E) = n$ elements, to each of which is assigned a probability $1/N$, then $P(E)$, the probability that E will occur, is the sum of the probabilities assigned to the individual outcomes. This is

$$P(E) = \frac{1}{N} + \frac{1}{N} + \cdots + \frac{1}{N} = \frac{n}{N}.$$

More generally then, for any subset E of equally likely simple events, we can write

$$P(E) = \frac{n(E)}{n(S)} \qquad \text{for every } E \subseteq S. \tag{3.2.1}$$

(Note that a direct application of the relative frequency argument to the sample points of the event E will also give the same answer if the relative frequencies of occurrence of all of the individual sample points are the same.) In particular, from equation (3.2.1), first setting $E = \varnothing$, the empty set, and then setting $E = S$, we obtain

$$P(\varnothing) = 0 \qquad P(S) = 1.$$

(If the sample space were infinite, modified definitions would be needed. We shall not discuss this point, however.)

   Example. A single die is thrown and $S = \{1, 2, 3, 4, 5, 6\}$. If the die is fair (balanced) and all outcomes are equally likely, then[3]

$$P(1) = P(2) = P(3) = P(4) = P(5) = P(6) = \tfrac{1}{6}.$$

The event $E = \{1, 5, 6\}$ is said to occur if the number showing on the top face of the die is 1 or 5 or 6. The probability of this event, $P(E)$, is the sum of the probabilities of the outcomes 1, 5, and 6; that is, $P(E) = (\tfrac{1}{6}) + (\tfrac{1}{6}) + (\tfrac{1}{6}) = \tfrac{1}{2}$. Alternatively, we can express the probability as $n(E)/n(S) = \tfrac{3}{6} = \tfrac{1}{2}$.

   Comments. In Section 3.3 we shall extend the idea of assigning probabilities to events by defining the *probability function*.

   It must be emphasized that probabilities are always related to a sample space. It is incongruous to discuss probabilities unless a sample space is explicitly defined or at least implied in such a way that there is no ambiguity. For example, any statement of "probability of rain" would be meaningless unless the time and region under consideration were specified. Likewise,

---

[3] Strictly speaking we should use the notation $P(\{1\})$, $P(\{2\})$, and so on, here to denote the fact that we are referring to the subset $\{1\}$ of S. (The difference between 1 and $\{1\}$ was pointed out on page 34.) However, this would lead to a rather cumbersome notation which is not necessary in practice because probability statements are made *only* about subsets of the sample space. Thus we omit the curly brackets in probabilities throughout.

"the probability of drawing a spade" requires the definition of a sample space in the form of a statement about the manner of drawing and the composition of the deck from which the card is to be drawn. An alternative specification would be an explicit listing of all possible sample points, of course.

Exercises

1. "It happens only once in a million times" said officials of the Fair Grounds after the eight horses in Wednesday's ninth race finished in the order of their program listing. Find the actual probability of this event.
2. Find the probability that the serial number of a $1 bill will contain a 3 or a 7. (There are eight digits in a serial number.)
3. Team A has played team B 15 times during the season. Team A has won 10 times and team B has won 5 times. If we take the probability that A will win a game with B to be $\frac{2}{3}$, find the probability that A will win (against B)
   (a) the next three games.
   (b) two of the next three games.
   (c) at least two out of the next three games.
4. An experiment has two parts: A balanced die is thrown and then a letter is selected by chance from the set $\{T, H\}$ by tossing a fair coin.
   (a) Describe a sample space (using set notation) for this experiment such that the individual outcomes, each of which consists of a number from 1 to 6 and a letter from the given set, have the same probability of occurring.
   (b) What is the probability of getting an H and a 3? An even number? A T? An H and an even number? A T and an odd number?
5. In a special deck of cards there are six red and four black cards. The red cards have values 1, 2, 4, 5, 6, and 7, and the black cards have values 4, 6, 8, and 9. If the ten cards are mixed and then one card is drawn at random, what is the probability that it is
   (a) a red card?
   (b) a 9?
   (c) a card with an even number?
   (d) a red card or a card with an even number?
   (e) one of the following: 1, 2, 5, 7, 8, or 9?
   (f) one of the following: 4 or 6?
6. The probability of getting H (heads) with a certain biased coin is $\frac{2}{3}$. Find the probability of getting
   (a) one H and one T in two tosses.
   (b) three H's in three tosses.
   (c) one H and two T's in three tosses.

7. Three fair coins are tossed. Find the probability of getting
    (a) 2 heads and 1 tail.   (b) 1 head and 2 tails.   (c) 3 tails.

8. Four fair coins are tossed. Find the probability of getting
    (a) 4 heads.   (c) 3 heads and 1 tail.
    (b) 2 heads and 2 tails.   (d) at least 2 heads.

9. Two fair dice are thrown. Find the probability of getting a point total of
    (a) 7.   (c) either 7 or 11.   (e) either 2 or 12.
    (b) 11.   (d) at most 4.

10. Find the probability that a card drawn from an ordinary 52-card deck will be
    (a) a face card (jack, queen, or king).
    (b) a face card or an ace.
    (c) a spade, an ace, or a 10.
    (d) *not* smaller than 10.

11. $E_1, E_2, \ldots, E_n$ are events in S. E_1' is the complement of the set E_1 relative to S, and so on. Show that the sets $\{E_1 \cup E_2 \cup \cdots \cup E_n\}$ and $\{E_1' \cap E_2' \cap \cdots \cap E_n'\}$ are disjoint (and hence that the corresponding events are mutually exclusive); show that their union is S, and therefore that $P(E_1 \cup E_2 \cup \cdots \cup E_n) = 1 - P(E_1' \cap E_2' \cap \cdots \cap E_n')$.

12. The eight letters S T A N D A R D are written separately on slips of paper and four slips are drawn at random without replacement. Find the probability that the letters drawn will spell S T A R. (It is not necessary to draw the letters in the proper order.)

13. There are six different ways—(6, 2, 1), (5, 3, 1), (5, 2, 2), (4, 3, 2), (4, 4, 1), and (3, 3, 3)—that three numbers below 6 can add to 9 and six different ways—(6, 3, 1), (6, 2, 2), (5, 4, 1), (5, 3, 2), (4, 4, 2), and (4, 3, 3)—that three numbers below 6 can add to 10. However, if three fair dice are tossed, the total 10 appears more often. Why?

14. A balanced coin is tossed twice and the game is won if H appears at least once. Comment on the following. Since there are four possibilities: HH, HT, TH, and TT, the probability of a win is $\frac{3}{4}$, TT being the only loser. On the other hand, there are really only three possibilities: H, TH, and TT, because if H is the first result there is no need to toss again. Thus the probability of a win is $\frac{2}{3}$.

15. Urn A contains 4 white, 5 red, and 6 black balls. Urn B contains 5 white, 6 red, and 7 black balls. One ball is selected from each urn. What is the probability that the two balls chosen will be of the same color?

16. About 40% of Americans have blood type A, 10% have type B, 5% have type AB, and 45% have type O. Find the probability that a couple marrying will have the same type. Assume that the proportions of blood types among men and women are the same as for people in general.

17. A woman at a supermarket learned that her bill for a cartfull of groceries was an exact dollar amount (no cents) and said, "I'll bet that doesn't happen very often." The checkout girl said, "Oh, that happens about once in a hundred times." How would the checkout girl do in a probability course?

3.3. Probability Function

At this point we know how to describe a sample space for a random phenomenon and we know methods of assigning probabilities to the individual sample points. We know, also, what an "event" is and how the probability of an event is determined for a finite sample space. Although many problems allow us to choose a *finite* sample space with *equally likely* sample points, there are many other problems in which the sample space is finite but the sample points *cannot* be equally likely, or in which the sample space is infinite. As illustrations, consider these examples.

   Example 1.   A biased coin is tossed. A sample space consists of the (finite) set of two sample points H and T which are not equally likely. (Also see Example 3 on page 42.)

   Example 2.   A coin is to be tossed until a head appears for the first time. A sample space S is the set of all possible outcomes: $\{H, TH, TTH, \ldots\}$. The space S clearly has an infinite number of points.

   To cope with such problems we need the concept of a probability function defined over a sample space that is not restricted to a finite number of equally likely sample points. We now define such a general probability function as the first step in formulating a mathematical system of probability. Suppose that, for a given random phenomenon, there is designated a sample space S with events $A_1, A_2, \ldots, A_k, \ldots$, which include $\varnothing$ and S. We define a probability function as follows.

   Definition.   With each event A_i we associate a real number $P(A_i)$, called the probability of A_i, with the following properties.

   1. $0 \le P(A_i) \le 1$.
   2. $P(S) = 1$.
   3. $P(A_i \cup A_j) = P(A_i) + P(A_j)$, if $A_i \cap A_j = \varnothing$ (that is, if A_i and A_j are mutually exclusive events).

The domain of the function P (that is, the world to which P applies) is the set of subsets of S, and the range of P (that is, the set of possible values of P) is the set of real numbers between zero and one inclusive. It should be noted that we do not say that the sample space must have equally likely sample points nor that it must be finite.

The mathematical theory does not require any particular assignment of probabilities $P(A_i)$ to the events A_i. Any of the classical, relative frequency, or subjective methods, or any other method of assigning probabilities, may be employed. The choice of probabilities in a specific situation could (theoretically) be contrary to experience (although it preferably should not be). The only theoretical requirements are that the three mathematical properties above be satisfied.

For example, for the experiment of tossing a fair coin with two outcomes, heads and tails, the probability of heads may be any value p and the probability of tails may be any value q, with the only restriction being that neither p nor q be negative and that $p + q = 1$. Experience with a fair coin would suggest the values $p = q = \frac{1}{2}$ as appropriate. If we ignored practical experience and took $p = 1$ and $q = 0$, we would have a perfectly proper probability function, but, for a fair coin, the theoretical framework would not conform to results obtained empirically in an actual succession of experiments of coin tossing.

Note that our definition of a probability function is consistent with both the classical and relative frequency methods of assigning probabilities to individual outcomes.

Consider property 1 of the definition. We have already observed in Section 1.4 that the probability of an outcome is a number between zero and one inclusive. This is also true from our definition of a probability function, for each outcome is a subset of S and consequently is an event, a single-member event in fact. The subset of S that represents this event is the subset that has this single outcome as its only element.

It was also pointed out in Section 1.4 that the sum of probabilities of all possible outcomes must be one. This is consistent with property 2 here, for S is the union of all the disjoint subsets representing the single-member events.

Regarding property 3, we note that the union of two mutually exclusive events, A_i and A_j, is also the union of the single-member events comprising A_i and the single-member events comprising A_j, and, since A_i and A_j have no single-member events in common, the probability of this union is simply the sum of the probabilities assigned to all these single-member events or outcomes.

Several laws of probability follow as theorems from this definition.

Theorem 1. $P(\varnothing) = 0$.

Proof. For any event A, $A \cup \varnothing = A$, and since A and $\varnothing$ are mutually exclusive,

$$P(A \cup \varnothing) = P(A) + P(\varnothing) = P(A).$$

Therefore, $P(\varnothing) = 0$.

Theorem 2. $P(A') = 1 - P(A)$, where $A' = S - A$ is the complement of A.

Proof. Since $A \cup A' = S$, and since A and A' are mutually exclusive,

$$P(A \cup A') = P(A) + P(A') = P(S) = 1.$$

Therefore, $P(A') = 1 - P(A)$.

Theorem 3. $P(A \cup B) = P(A) + P(B) - P(A \cap B)$.

Proof. Since $A \cup B = A \cup (B \cap A')$ and $B = (A \cap B) \cup (B \cap A')$, $P(A \cup B) = P(A) + P(B \cap A')$ and $P(B) = P(A \cap B) + P(B \cap A')$. Subtracting, $P(A \cup B) - P(B) = P(A) - P(A \cap B)$, or $P(A \cup B) = P(A) + P(B) - P(A \cap B)$. [*Note*: A and $B \cap A'$ are mutually exclusive, etc.]

Example. To illustrate Theorem 3, let

$A =$ the event that a card drawn from a bridge deck will be a spade,

$B =$ the event that a card drawn from a bridge deck will be an ace,

so

$A \cap B =$ the event that the ace of spades is drawn.

Since all cards are equally likely to be drawn,

$$P(A) = \tfrac{13}{52} = \tfrac{1}{4}, \quad P(B) = \tfrac{4}{52} = \tfrac{1}{13}, \quad P(A \cap B) = \tfrac{1}{52}.$$

Then

$$P(A \cup B) = \tfrac{1}{4} + \tfrac{1}{13} - \tfrac{1}{52} = \tfrac{16}{52} = \tfrac{4}{13}.$$

This result is easily verified, for, since we have equally likely outcomes,

$$P(A \cup B) = n(A \cup B)/n(S) = (13 \text{ spades} + 4 \text{ aces} - \text{ace of spades})/52,$$

the ace of spades having been counted twice, first with the spades and again with the aces.

This result can be extended to the union of any number of events. For three events, for example,

$$P(A \cup B \cup C) = P(A) + P(B) + P(C)$$
$$- P(A \cap B) - P(A \cap C) - P(B \cap C)$$
$$+ P(A \cap B \cap C).$$

The reader should draw a Venn diagram to illustrate this result. The formula when there are k events is not hard to see; the reader may wish to attempt to write it down. (It is given in Section 3.7 as a check.)

Exercises

1. One card is drawn from a bridge deck. Find the probability that
   (a) it is either a face card or a red card.
   (b) it is neither a face card nor a red card.
2. A box contains 6 red, 8 white, and 10 blue balls. One ball is taken from the box. Find the probability that the ball is
   (a) white.   (b) not white.   (c) either white or red.
3. Prove that if $A \subseteq B$, then $P(A) \le P(B)$. (*Hint:* Express B as the union of two mutually exclusive events.)
4. Prove that, for any two events A and B, $P(A \cup B) \le P(A) + P(B)$.
5. A card is drawn at random from a deck of 52 cards. What is the probability that it is
   (a) a club?   (b) an ace?   (c) the ace of clubs?
6. What is the probability of obtaining a total of 10 on a single simultaneous toss of two dice? What is the probability of getting at least one 10 if the two dice are thrown twice?
7. A secretary writes three letters and addresses the corresponding envelopes. She then absent-mindedly places the letters in the envelopes at random. What is the probability that each letter is in the proper envelope? What is the probability that *no* letter is in the proper envelope?
8. In a roomful of 25 people, what is the probability that two or more people have the same birthday? (Assume 365 days in a year. *Hint:* Find the probability of the complementary event.)
9. A piece of equipment has three components, each of which must be operating properly for the equipment to function. The probabilities that the components operate properly during a certain time interval are, respectively, p_1, p_2, and p_3. Assume that the components operate properly or do not operate properly quite independently of one another. Find the probability that, during the given time interval, the piece of equipment will cease to function.
10. What is the probability of holding at least one long suit (five or more cards of a suit) at bridge?
11. What is the probability of holding at least one singleton at bridge (that is, only one card of a suit)?
12. What is the probability of holding n assigned cards at bridge?
13. (a) Prove that $P(A \cup B) = P(A) + P(B) - P(A \cap B)$.
    (b) Prove the extended result beginning $P(A \cup B \cup C) = P(A) + \cdots$.
    (c) Two ordinary dice each with faces $1, \ldots, 6$ are thrown together.

What is the probability of obtaining any pair *or* a total of 4 with one throw?

14. An unreliable fellow student asks you to solve for him the following problem: $P(A) = \frac{2}{3}$, $P(A \cap B) = \frac{1}{6}$, $P(B) = \frac{11}{18}$; what is $P(A \cup B)$? What answer would you give him?

15. If $P(A') = 0.6$, $P(B) = 0.5$, and $P(A \cup B)' = 0.4$, find
   (a) $P(A \cap B)$.   (b) $P(A' \cup B')$.

16. The probability that a hurricane will hit a certain Florida city in June is 0.04; the probability that a hurricane will hit the city in September is 0.06; the probability that a hurricane will hit both in June and in September is 0.01. What is the probability that the city will not be hit by a hurricane in either month?

17. What is the probability of getting two, three, or four aces in a hand of five cards drawn at random without replacement from an ordinary bridge deck?

18. There are six married couples at a dinner. Place cards have simply Mr. L, Mrs. B, and so on. What is the probability that no two couples have the same last initial? State what assumptions you are making.

19. From a group of N elements (all equally likely to be selected), n are chosen at random, with replacement. Find the probability that at least one element is chosen more than once.

3.4. Conditional Probability

The idea of conditional probability is an extremely important one. The following example illustrates the concept.

Example. Suppose that from a class of 15 boys and 10 girls a student is selected by chance to perform a certain task during Monday's class period and a *different* student is selected by chance to perform the same task during Wednesday's class period. *Given the result of Monday's selection*, what is the probability that on Wednesday a boy will be selected? Two answers are possible according to what happened on Monday:

1. Given that Monday's task fell to a boy, the probability that a boy will be selected on Wednesday is $\frac{14}{24}$, for there are 14 boys different from the one selected on Monday and there is a total of 24 students who were not selected on Monday.

2. Given that a girl was selected on Monday, the probability that a boy will be selected on Wednesday is $\frac{15}{24}$, for there are now 15 boys not selected on Monday and there is (again) a total of 24 students who were not selected on Monday.

Note that if no information about the result of Monday's drawing is given, the probability that a boy will be selected on Wednesday is not a conditional probability in the sense above. Intuitively we conclude that this probability should be $\frac{15}{25}$. (That this is indeed correct will be shown later.)

To summarize, we have three different questions with three different answers:

1. Given that Monday's task fell to a boy, what is the probability that a boy will be selected on Wednesday? The answer is $\frac{14}{24}$.

2. Given that Monday's task fell to a girl, what is the probability that a boy will be selected on Wednesday? The answer is $\frac{15}{24}$.

3. Given no information about Monday, what is the probability that a boy will be selected on Wednesday? The answer is $\frac{15}{25}$.

The first two are conditional probabilities; the third is not. The probabilities that a girl will be selected on Wednesday can be evaluated in similar fashion.

The general notation for the "conditional probability of B given that event A has happened" is $P(B|A)$. It is read "the probability of B given A."

Example. For our example, we can denote by B the event that a boy is selected on Monday; we also write G for the event that a girl is selected on Monday and C for the event that a boy is selected on Wednesday. It follows that

$$P(C|B) = \tfrac{14}{24} \qquad P(C|G) = \tfrac{15}{24}.$$

A tree diagram may be used to indicate the conditional probabilities that arise from a problem.

Example. For our example, an appropriate tree diagram is given in Figure 3.1. We see from Figure 3.1 that there are four possible events on Wednesday. The four corresponding conditional probabilities are given on the last set of four branches.

FORMULA FOR CONDITIONAL PROBABILITY. In general, when we are given that a particular event has already occurred, the sample space for a subsequent event is reduced *to those outcomes that are possible in the light of this information.* We then take this reduced sample space into account in calculating probabilities. This can be illustrated by a Venn diagram. Suppose two events A and B are *not* mutually exclusive and we want the probability $P(B|A)$. [When A and B *are* mutually exclusive, $P(A|B) = P(B|A) = 0$.] Before we have the information that A has occurred, we can represent the sample space by the rectangle S in Figure 3.2(a). When we are

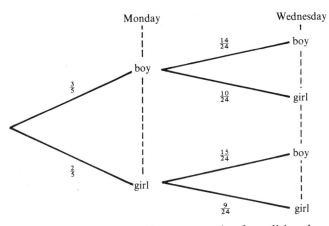

Figure 3.1.   Tree diagram for an example of conditional probabilities.

told that A has occurred, we can reduce the conditional sample space to just those outcomes which belong to A [Figure 3.2(b)]. Thus we define the conditional probability of B given A to be

$$P(B|A) = P(A \cap B)/P(A) \qquad \text{for } P(A) \neq 0$$

or

$$P(B|A) \cdot P(A) = P(A \cap B).$$

This formula can easily be justified in the case of a finite sample space with equally likely outcomes, for

$$P(B|A) = \frac{n(A \cap B)}{n(A)} = \frac{n(A \cap B)/n(S)}{n(A)/n(S)} = \frac{P(A \cap B)}{P(A)}.$$

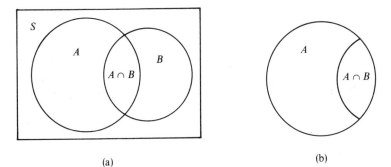

(a)   (b)

Figure 3.2.   Reducing the sample space for evaluating conditional probabilities.

It is also true that

$$P(A|B) \cdot P(B) = P(A \cap B).$$

Since $P(B) = P(B|S)$ and $P(S) = 1$, we can write as a special case of the formula above

$$P(B|S) = P(B \cap S)/P(S).$$

Note. In a sense, *all* probabilities are conditional probabilities. For in order to speak about a probability at all, we must have a sample space S. Thus it would be more correct, strictly speaking, to write $P(A|S)$ rather than $P(A)$. Since the existence of S is always understood, however, we do not need to use this more complicated notation.

Example. Two cards are drawn without replacement from a bridge deck (that is, the first card is not replaced before the second card is drawn). What is the probability that the second card is a spade, given that the first was a spade? Let A and B be the events that the first card is a spade and the second card is a spade, respectively; then we want the probability $P(B|A)$. By the formula above we have that

$$P(B|A) = P(B \cap A)/P(A)$$

$$= \frac{\frac{13}{52} \cdot \frac{12}{51}}{\frac{13}{52}} = \frac{12}{51} = \frac{4}{17}.$$

This is easily verified, for if the first card was a spade, there are 12 spades left among the remaining 51 cards for the second drawing. The original sample space S for this example is the set of the $52 \cdot 51 = 2652$ ways of drawing two cards in succession and without replacement from a deck of 52 cards. The event A is the subset of the $13 \cdot 51 = 663$ ways of drawing a spade and then another card (which may also be a spade). The event B is the subset of the $51 \cdot 13 = 663$ ways of drawing a spade second and any other card first. (Although A and B have the same number of elements, they are not the same set.) The event $A \cap B$ is the subset of the $13 \cdot 12 = 156$ ways of drawing two spades in succession and is, of course, a subset of A and also a subset of B. Thus in Figure 3.2, S has 2652 elements, A and B each have 663 elements, and $A \cap B$ has 156 elements. The probabilities are

$$P(A) = P(B) = \tfrac{663}{2652} = \tfrac{1}{4},$$

$$P(A \cap B) = \tfrac{156}{2652} = \tfrac{1}{17},$$

$$P(B|A) = \tfrac{156}{663} = \tfrac{4}{17}.$$

These agree with those obtained above by the formula.

Note that the event "the first card is a spade *given* that the second is a spade" is not the same as the event "the first card is a spade *and* the second is a spade." Although precisely the same outcomes belong to both events, the sample spaces are different, as indicated in Figure 3.2. $P(B|A)$ is the probability that the event B occurs, when we take into consideration the given information about the event A, that is, when the sample space is the set A. $P(A \cap B)$, on the other hand, is the probability that both A and B occur when the sample space is S. $P(B|A)$ is just one factor of the formula for $P(A \cap B)$, the other factor being $P(A)$, a number never greater than one, so that $P(B|A)$ is generally larger than $P(A \cap B)$.

For more than two events, the appropriate formula for conditional probabilities becomes more complicated but follows a definite pattern. For example,

$$P(A \cap B \cap C) = P(A) \cdot P(B|A) \cdot P(C|B \cap A),$$

or we can write any one of the other five expressions obtained by interchanging the roles of A, B, and C.

Example. Consider the experiment of drawing three cards without replacement from a bridge deck, and the events A = first card is a spade, B = second card is a spade, C = third card is a spade. Then

$$P(A \cap B \cap C) = \left(\tfrac{13}{52}\right) \cdot \left(\tfrac{12}{51}\right) \cdot \left(\tfrac{11}{50}\right) = \tfrac{11}{850},$$

according to the formula above.

The extended conditional probability formula for k events is given in the summary at the end of the chapter. This general formula, and the formulas above, are valid whether the sample space is finite or infinite, countable or noncountable, whether the individual outcomes are equally likely or not, and whether the events occur at the same time or in succession.

Exercises

1. One card is drawn from a standard bridge deck. Find the probability that it is
    (a) a spade or a heart.
    (b) a spade or an ace.
    (c) a spade, given that it is black.
2. Two cards are drawn with replacement from a bridge deck. Find the probability that
    (a) both are kings.
    (b) the second is a ten, given that the first is a face card.
    (c) the second is a ten, given that the first is a ten.
    Repeat the problem for cards drawn without replacement.

3. The odds against a certain basketball team getting a bid to the NCAA tournament are 7:1. The odds against getting a bid to the NIT are 4:1. What are the corresponding probabilities that the team will get these bids? If the odds against getting bids to both tournaments are 11:1, find the probability that the team will not get a bid to either tournament.

4. If two cards are drawn from a bridge deck without replacement (that is, the first card is not replaced before the second is drawn), find the probability that
   (a) both are spades.
   (b) the first is a spade and the second is a heart.
   (c) the first is a spade and the second is an ace.
   (d) the second is a spade, given that the first is a spade.
   (e) the second is a spade, given that the first is black.
   (f) one is a spade and the other is an ace.

5. Find the conditional probability that a tossed die shows a number greater than 3, given that the number shown is even.

6. Find the conditional probability that the sum obtained when tossing two dice simultaneously is at least 9, if at least one of the dice shows a 5 or smaller number.

7. Four coins are tossed simultaneously. Find the probability of getting exactly three heads, if we are informed that at least one coin is a "head."

8. A bag contains three white marbles and seven red marbles. Two marbles are drawn at random from the bag. Find the probability that both are red, if the first marble is
   (a) replaced before the second is drawn.
   (b) not replaced before the second is drawn.

9. A hat contains 11 tickets numbered from one to eleven. Two tickets are drawn without replacement. What is the probability that
   (a) the first ticket has an even number and the second an odd number?
   (b) one ticket has an odd number and the other has an even number?
   (c) the second ticket has an even number, given that the first has an odd number?

10. Five integers are chosen at random· What is the probability that
    (a) exactly four of them are even?
    (b) at least three of them are odd?

11. Two cards are drawn without replacement from a bridge deck (that is, the first is not replaced before the second is drawn). Find the probability that
    (a) both are kings.
    (b) the first is a heart and the second is black.
    (c) the first is a heart and the second is red.

(d) the first is a heart and the second is an ace.

(e) the first is a king and the second is an ace.

12. A group of students consists of 300 boys and 200 girls. One-third of the boys are freshmen and the remainder are sophomores. One-fourth of the girls are freshmen and the remainder are sophomores. A student is selected at random from the whole group. What is the probability that the student is

    (a) a freshman girl?

    (b) a freshman?

    (c) a freshman, given that the student is a girl?

    (d) a girl, given that the student is a freshman?

13. $P(A) = 0.30$, $P(B|A) = 0.56$, and $P(B) = 0.70$. Find

    (a) $P(A|B)$.   (b) $P(A \cup B)$.

14. Events A and B are mutually exclusive. $P(A) = 0.35$, $P(B) = 0.60$. Find

    (a) $P(A')$.   (b) $P(A \cup B)$.   (c) $P(A|B)$.

15. $P(A) = 0.6$, $P(B|A) = 0.2$, $P(B) = 0.3$. Find

    (a) $P(A \cap B)$.   (b) $P(A \cup B)$.   (c) $P(A|B)$.

16. $P(A) = 0.4$, $P(B) = 0.2$, and $P(B|A) = 0.3$. Find

    (a) $P(A \cup B)$.   (b) $P(A|B)$.   (c) $P(A' \cap B')$.

17. If $P(A) = 0.4$, $P(B) = 0.8$, and $P(B|A) = 0.6$, find

    (a) $P(A|B)$.   (b) $P(A \cup B)$.

18. $P(A) = 0.4$, $P(B) = 0.5$, $P(A|B) = 0.2$.

    (a) Are A and B mutually exclusive?

    (b) Find $P(A \cap B)$.

    (c) Find $P(A')$.

    (d) Find $P(A \cup B)$.

19. If $P(A) = 0.3$, $P(B) = 0.5$, and $P(A|B) = 0.4$, find

    (a) $P(A \cup B)$.   (b) $P(B|A)$.   (c) $P(A' \cup B')$.

20. If $P(A) = 0.6$, $P(B) = 0.2$, and $P(A \cap B) = 0.1$, find

    (a) $P(A|B)$.   (c) $P(A \cup B)$.   (e) $P(\emptyset)$.

    (b) $P(B|A)$.   (d) $P(S)$.

Are A and B mutually exclusive?

21. A box contains 25 red balls, 25 blue balls, and 50 white balls, thoroughly mixed. Two balls are drawn from the box with replacement. Find the probability that

    (a) at least one is red.   (b) both are the same color.

If the balls are drawn without replacement, find the probability that

    (c) the first is red.

    (d) both are red.

    (e) the second is blue given that the first is white.

22. The probability that, in winter, a certain locality will have "cold weather" (below 20°F) is 0.6. The probability that it will have "skiing snow"

(6 inches or more) is 0.7. The probability that it will have both cold weather and skiing snow is 0.4. Find the probability of skiing snow given cold weather, and the probability of cold weather given skiing snow.

23. Find the probability that the sum of the numbers showing on two dice is 8, given that at least one die is not a five.

24. Find the probability that the sum obtained when tossing two dice is either 7 or 11, given that at least one die is not a six.

25. Let S be the set of all students in class, A the set of all students in class with first initial J, and B the set of all students in class between 19 and 20 years of age. Suppose that there are 75 students in class, that 17 of these have first initial J, that 23 are between 19 and 20 years of age, and that 5 fall into both categories. Find the probability of A given B, and describe in words the students belonging to this event.

26. At a meeting of a college student organization there are 6 seniors, 18 juniors, 36 sophomores, and 12 freshmen. Of the 72 present, 48 are men and 24 are women. If one member is selected at random as a delegate to a convention and then a different member is selected at random as an alternate, find the probability that
    (a) the delegate is a man and the alternate is a woman.
    (b) both are freshmen.
    (c) both are women.
    (d) the alternate is a man given that the delegate is a man.
    (e) the delegate is a senior given that the alternate is a sophomore.
    (f) either the delegate or the alternate, or both, are juniors.
    (g) one is a senior and the other is a freshman.
    (h) the delegate is not a junior.

3.5. Theorem of Total Probability and Bayes' Theorem

Example. To introduce and illustrate the material of this section, we consider the following problem. Jim tells Dave at school one morning that he has bought a new sweater, and he asks Dave to guess what color he bought. Dave believes that it is either red or gold, but he is not sure which of these colors to choose. However, he has several pieces of information that might help. He knows there are only four stores that handles sweaters of the kind Jim likes; we denote these by B_1, B_2, B_3, and B_4. Moreover, from his own shopping and talking to salesmen, Dave knows that the four stores sell red sweaters in the percentages 25, 50, 30, and 50, respectively, and gold sweaters in percentages 40, 30, 45, and 30, respectively. (Thus 25% of all sweaters sold by store B_1 are red, 30% of all sweaters sold by store B_2 are gold, and so on.) Dave also knows that Jim buys 40% of his clothes at B_1, 25% at B_2, 20% at B_3, and 15% at B_4. How can Dave use this information?

First let us convert the percentages to probabilities. If we denote the events that the sweater is red or gold by A_1 and A_2, respectively, and also use the store symbols B_1, B_2, B_3, and B_4 to denote the events that Jim bought the sweater (whatever its color) at stores B_1, B_2, B_3, and B_4, respectively, we can write

$$P(A_1|B_1) = 0.25, \quad P(A_1|B_2) = 0.50, \quad P(A_1|B_3) = 0.30, \quad P(A_1|B_4) = 0.50,$$

$$P(A_2|B_1) = 0.40, \quad P(A_2|B_2) = 0.30, \quad P(A_2|B_3) = 0.45, \quad P(A_2|B_4) = 0.30,$$

while the probabilities that Jim bought his sweater in the various stores are

$$P(B_1) = 0.40, \quad P(B_2) = 0.25, \quad P(B_3) = 0.20, \quad P(B_4) = 0.15.$$

Now the event that Jim bought a red sweater can be written as the union of the mutually exclusive events that he bought a red sweater at B_1, or at B_2, or at B_3, or at B_4. Thus, since A_1 denotes the event that Jim bought a red sweater, we can write

$$A_1 = (B_1 \cap A_1) \cup (B_2 \cap A_1) \cup (B_3 \cap A_1) \cup (B_4 \cap A_1).$$

The probability $P(A_1)$ that Jim bought a red sweater is thus the sum of the probabilities of the four mutually exclusive events $B_i \cap A_1$, for $i = 1, 2, 3, 4$; that is,

$$P(A_1) = P(B_1 \cap A_1) + P(B_2 \cap A_1) + P(B_3 \cap A_1) + P(B_4 \cap A_1).$$

Each of the probabilities on the right can be expressed as the product of two probabilities: (1) that Jim bought his sweater at, say, B_i, and (2) *given* that he bought it at B_i, that it was red (that is, event A_1 occurred). Substituting for

$$P(B_i \cap A_1) = P(B_i) \cdot P(A_1|B_i),$$

we obtain the formula

$$P(A_1) = P(B_1)P(A_1|B_1) + P(B_2)P(A_1|B_2) + P(B_3)P(A_1|B_3) + P(B_4)P(A_1|B_4).$$

Putting in the numerical values above provides

$$P(A_1) = (0.40)(0.25) + (0.25)(0.50) + (0.20)(0.30) + (0.15)(0.50)$$
$$= 0.360.$$

By an exactly similar argument we can find the probability that the sweater is gold (event A_2) as

$$P(A_2) = (0.40)(0.40) + (0.25)(0.30) + (0.20)(0.45) + (0.15)(0.30) = 0.370.$$

Thus, very slightly, the odds favor the conclusion that the sweater is gold rather than red, by 0.370 to 0.360. (Note, also, that there is a probability of $1 - 0.360 - 0.370 = 0.270$ that the sweater is neither red nor gold.)

The foregoing method of obtaining the probabilities for red and gold relies on the following result.

THEOREM OF TOTAL PROBABILITY.   Suppose we can partition the sample space S into k disjoint sets (events), say $B_1, B_2, \ldots, B_k$, so that $B_1 \cup B_2 \cup \cdots \cup B_k = S$. (This is illustrated in Figure 3.3 for $k = 4$.) Let A be an event belonging to S. Then the probability of the event A is given by

$$P(A) = P(A \cap B_1) + P(A \cap B_2) + \cdots + P(A \cap B_k).$$

This is the theorem of total probability.

(In words, if an event can occur in several different ways that are mutually exclusive, the probability of the event is equal to the sum of the probabilities of the mutually exclusive ways in which it can occur. We have already seen a simplified version of this theorem stated as property 3 of the definition of the probability function in Section 3.3.)

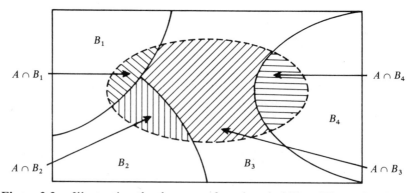

Figure 3.3.   Illustrating the theorem of total probability. (The entire shaded region is A; the rectangle is S.)

Proof. Since the sets $B_1, B_2, \ldots, B_k$ are disjoint, the intersections $A \cap B_1, A \cap B_2, \ldots, A \cap B_k$ are disjoint as well (see Figure 3.3). Since the union of the *B*'s is *S* and since *A* is a subset of *S*, we can express *A* as the union of these *k* disjoint sets; thus

$$A = (A \cap B_1) \cup (A \cap B_2) \cup \cdots \cup (A \cap B_k).$$

It follows that

$$P(A) = P(A \cap B_1) + P(A \cap B_2) + \cdots + P(A \cap B_k).$$

A FURTHER DEVELOPMENT. If we know the conditional probabilities $P(A|B_j)$ for all $j = 1, 2, \ldots, k$, and we also know the probabilities $P(B_j)$ for all j, the probability of the event *A* can be written in another useful form, as already used in the example. According to the definition of conditional probability, we can write

$$P(A \cap B_j) = P(A|B_j) \cdot P(B_j).$$

Hence

$$P(A) = P(A|B_1) \cdot P(B_1) + \cdots + P(A|B_k) \cdot P(B_k)$$

$$= \sum_{j=1}^{k} P(A|B_j) \cdot P(B_j).$$

CONTINUATION OF THE EXAMPLE. Let us now assume that Dave has guessed the color of Jim's new sweater as gold (A_2) and that this is correct; that is, we know that event A_2 has occurred. Jim now asks Dave to guess where the sweater was purchased. How can Dave proceed? What he can do is to find the probability that the sweater was bought at each of the stores given that it was gold—that is, find $P(B_i|A_2)$ for each $i = 1, 2, 3, 4$, and choose, as his guess, the store B_i for which $P(B_i|A_2)$ is largest. Now, by the rules of conditional probability,

$$P(B_i \cap A_2) = P(B_i|A_2) \cdot P(A_2) = P(A_2|B_i) \cdot P(B_i).$$

Thus we can write

$$P(B_i|A_2) = \frac{P(A_2|B_i) \cdot P(B_i)}{P(A_2)}.$$

For $i = 1$, therefore, substituting from above,

$$P(B_1|A_2) = \frac{(0.40)(0.40)}{0.370} = \frac{160}{370}.$$

Similar calculations give

$$P(B_2|A_2) = \frac{75}{370}, \quad P(B_3|A_2) = \frac{90}{370}, \quad P(B_4|A_2) = \frac{45}{370}.$$

The most likely store, then, is B_1, the second most likely is B_3, then B_2, then B_4. So Dave should guess B_1. What we have done in this calculation is to apply Bayes' theorem, which we now set out more formally.

BAYES' THEOREM.   Suppose we have the situation in the foregoing theorem where the sample space S is partitioned into k mutually exclusive sets (events) $B_1, B_2, \ldots, B_k$, so that $B_1 \cup B_2 \cup \cdots \cup B_k = S$. (This is illustrated in Figure 3.3 for $k = 4$.) Let A be an event belonging to S and suppose we want the probability of a particular event B_i *given A*. In symbols, we want the conditional probability $P(B_i|A)$. According to the definition of conditional probability,

$$P(B_i|A) = \frac{P(B_i \cap A)}{P(A)}.$$

Both the numerator and the denominator of this fraction can be written in other forms using the results given above, so that

$$P(B_i|A) = \frac{P(A|B_i) \cdot P(B_i)}{\sum_{j=1}^{k} P(A|B_j) \cdot P(B_j)}.$$

This is the formal statement of *Bayes' theorem*.

Example.   Suppose there are three boxes with contents as shown in the second and third columns of the following table.

Box No.	No. of marbles		Chance of drawing white marble	Chance of choosing box
	Red	White		
1	5	5	$\frac{1}{2} = P(A|B_1)$	$\frac{1}{4} = P(B_1)$
2	5	10	$\frac{2}{3} = P(A|B_2)$	$\frac{1}{4} = P(B_2)$
3	10	5	$\frac{1}{3} = P(A|B_3)$	$\frac{1}{2} = P(B_3)$

The figures in the fourth column are easy to obtain. The fifth-column figures are obtained by assuming that one box is chosen according to the following scheme: Two fair coins are tossed. If both are heads (probability $\frac{1}{4}$), box 1 is chosen; if both are tails (probability $\frac{1}{4}$), box 2 is chosen; otherwise (probability $\frac{1}{2}$), box 3 is chosen. Suppose a box is selected and one marble is drawn from the selected box.

1. What is the probability that it is white?

2. If it is white, what is the probability that it was drawn from box 1?

To answer the first question we appeal to the theorem of total probability; to answer the second question we apply Bayes' theorem, as follows.

1. The sample space S is the set of all possible outcomes. It can be partitioned into three mutually exclusive sets corresponding to the three boxes, with points representing red and white marbles in each set. The outcome "white" can occur with any one of the three events "box 1," "box 2," or "box 3." Thus we shall denote by A the event that the marble is white, and by B_1, B_2, and B_3 the events that box 1, box 2, and box 3 was chosen, respectively. The probability of drawing a white marble is then given by

$$P(A) = P(A \cap B_1) + P(A \cap B_2) + P(A \cap B_3)$$
$$= P(A|B_1)P(B_1) + P(A|B_2)P(B_2) + P(A|B_3)P(B_3)$$
$$= \tfrac{1}{2} \cdot \tfrac{1}{4} + \tfrac{2}{3} \cdot \tfrac{1}{4} + \tfrac{1}{3} \cdot \tfrac{1}{2} = \tfrac{11}{24}.$$

2. The question asks for $P(B_1|A)$. Applying Bayes' theorem, we find that

$$P(B_1|A) = \frac{P(A|B_1)P(B_1)}{P(A|B_1)P(B_1) + P(A|B_2)P(B_2) + P(A|B_3)P(B_3)}$$

$$= \frac{\tfrac{1}{2} \cdot \tfrac{1}{4}}{\tfrac{1}{2} \cdot \tfrac{1}{4} + \tfrac{2}{3} \cdot \tfrac{1}{4} + \tfrac{1}{3} \cdot \tfrac{1}{2}} = \frac{3}{11}.$$

[Note that $P(B_2|A) = \tfrac{4}{11} = P(B_3|A)$, and the three probabilities $P(B_i|A)$, $i = 1, 2, 3$, add to one, as they should.]

ADDITIONAL EXAMPLES ILLUSTRATING BAYES' THEOREM.

One interpretation of Bayes' theorem occurs in connection with causes and effects. If A is an effect that can result only after one of the causes B_j, then Bayes' theorem gives the probability of a particular cause B_i, given the effect A. Examples of this interpretation are the following.

Example 1. An electronic device has three components and the failure of any one of them may, or may not, cause the device to shut off automatically. Furthermore, these failures are the only possible causes for a shutoff, and the probability that two of the components will fail simultaneously is negligible. At any time, component B_1 will fail with probability 0.1, component B_2 will fail with probability 0.3, and component B_3 will fail with probability 0.6. Also, if component B_1 fails, the device will shut off with probability 0.2; if component B_2 fails, the device will shut off with probability 0.5; if component B_3 fails, the device will shut off with probability 0.1. The device suddenly shuts off. What is the probability that the shutoff was caused by failure of component B_1?

The event that happened is the shutting off of the device, and the mutually exclusive events are the failures of B_1, B_2, and B_3. Let us designate these events by the letters A, B_1, B_2, and B_3, respectively. Then

$$P(B_1|A) = \frac{P(A|B_1)P(B_1)}{\sum\limits_{j=1}^{3} P(A|B_j)P(B_j)}$$

$$= \frac{(0.2)(0.1)}{(0.2)(0.1) + (0.5)(0.3) + (0.1)(0.6)} = \frac{2}{23}.$$

[Similarly, $P(B_2|A) = \frac{15}{23}$ and $P(B_3|A) = \frac{6}{23}$. Note that the three probabilities add to one, as they should.]

Example 2. Suppose that three drugs are known to have cured patients with a certain disease; that there is no other known method of cure; that the three drugs may be administered together but that they do not interact; and that, in a particular patient, only one of the drugs will effect a cure, an event we denote by A. Let us designate by B_1, B_2, and B_3 the events that the three drugs produce responses. Suppose that the corresponding probabilities that the drugs will produce a response in the patient are 0.2, 0.5, and 0.3, respectively, and the probabilities that the patient will be cured if the corresponding drug does produce a response are 0.1, 0.2, and 0.3, respectively. All three drugs are administered simultaneously to a patient and the patient is cured. What is the probability that drug B_1 effected the cure?

Applying Bayes' theorem we have

$$P(B_1|A) = \frac{(0.1)(0.2)}{(0.1)(0.2) + (0.2)(0.5) + (0.3)(0.3)} = \frac{2}{21}.$$

[Similarly, $P(B_2|A) = \frac{10}{21}$ and $P(B_3|A) = \frac{9}{21} = \frac{3}{7}$. Note that the three probabilities add to one, as they should.]

Exercises

1. If $B_1 \cup B_2 \cup B_3 \cup B_4 = S$, $P(A \cap B_1) = 0.15$, $P(A \cap B_2) = 0.25$, $P(A \cap B_3) = 0.20$, $P(A \cap B_4) = 0.10$, and $P(B_1) = P(B_2) = P(B_3) = P(B_4) = 0.25$, find $P(A)$.

2. If $F_1 \cup F_2 \cup F_3 = S$, $P(E|F_1) = 0.12$, $P(E|F_2) = 0.16$, $P(E|F_3) = 0.24$, $P(F_1) = 0.25$, $P(F_2) = 0.35$, and $P(F_3) = 0.40$, find $P(E)$.

3. An automobile dealer sells three types of cars—hard tops, convertibles, and station wagons—in the following percentages, respectively: 50, 15, and 35. Of the hard tops sold, 40% are red; of the convertibles sold, 70% are red; of the station wagons sold, 25% are red. Find the probability that a car sold will be red.

4. The proportion of patients in a hospital who are men, women, and children, respectively, are: 40, 35, and 25. Of the men patients, 50% are surgery patients; of the women, 40% are surgery patients; of the children, 30% are surgery patients. Find the probability that a patient chosen at random is a surgery patient.

5. If $P(A) = \frac{1}{3}$, $P(B|A) = \frac{1}{4}$, and $P(B'|A) = \frac{2}{3}$, find $P(B)$. (B' is the complement of B.)

6. A box contains five red and four blue chips and a second box contains three red and six blue chips. A chip is selected at random from the first box and placed in the second. Then one chip is drawn from the second box. What is the probability that the second chip is blue?

7. Assume that there are an equal number of male and female students in a high school and that the probability is 0.2 that a male student will be a science major and 0.05 that a female student will be a science major. Find the probability that a student selected at random will be
   (a) a male science major.
   (b) a science major.
   (c) male, given that the student is a science major.

8. Box I contains 7 red and 4 white balls; box II contains 3 red and 6 white balls. A box is selected by tossing a coin, box I corresponding to heads and box II to tails, and one ball is withdrawn from the box selected. If the ball selected is red, what are the probabilities that the coin was heads or tails? Check that your probabilities add to one.

9. Three boxes contain the following balls:

Box	Red	White
1	3	2
2	4	5
3	1	2

A box is selected at random and a ball is withdrawn. If the ball is red, what is the probability that it came from box 3?

10. Three machines produce the same kind of bolt; machine A produces 50% of the total output, but 10% of its output are defective; machine B produces 30% of the total output, but 8% are defective; machine C produces 20% of the total output, but 6% are defective. A bolt is selected from a bin in which the total output is stored and it is defective. What are the probabilities that it was produced by machine A, B, or C? Check that your probabilities add to one.

11. A certain man is a frequent visitor at the race track. When he uses his own "scientific" method he wins 10% of the time. When he follows the

suggestions of a tout he wins 40% of the time. When he follows his wife's suggestions he wins 80% of the time. Like many husbands, he uses his own method 60% of the time, he listens to the tout 30% of the time, and he listens to his wife 10% of the time. Today he won. What method do you think he used? (That is, what method is most probable?)

12. Four shelves of books in a professor's office have journals with different-colored covers as follows:

> Shelf 1: 6 red, 5 blue, 3 green, 2 yellow
> Shelf 2: 5 red, 7 blue, 6 green, 6 yellow
> Shelf 3: 4 red, 3 blue, 5 green, 3 yellow
> Shelf 4: 3 red, 6 blue, 4 green, 3 yellow

Usually the professor chooses a shelf and then selects a journal from it at random, but because the shelves are at different heights from the floor, he chooses shelf 1 20% of the time, shelf 2 40% of the time, shelf 3 30% of the time, and shelf 4 10% of the time. The Professor chooses a shelf and one journal is taken from it. It has a green cover. Find the probability that it came from shelf 3.

13. Stores X, Y, and Z sell brands A, B, and C of men's shirts. A customer buys 50% of his shirts at X, 20% at Y, and 30% at Z. Store X sells 25% brand A, 40% brand B, and 25% brand C. Store Y sells 40% brand A, 20% brand B, and 30% brand C. Store Z sells 20% brand A, 40% brand B, and 20% brand C. The customer comes home one day with a new shirt of brand C. What is the probability that it was purchased at store X?

14. Refer to the continuation of the "Jim's sweater" example on page 63. If the color of the sweater had been red, what are the respective probabilities that Jim purchased it from stores B_1, B_2, B_3, and B_4?

3.6. Independent Events

The concept of conditional probability leads to the notion of *independence* of events. When we are concerned with the probability of two or more events occurring, we shall often wish to know if the events affect each other or are independent. Formally we can define independence of two events A and B as follows.

Definition. Events A and B are independent if and only if

$$P(A \cap B) = P(A) \cdot P(B). \qquad (3.6.1)$$

An equivalent definition is that A and B are independent if and only if

$$P(A|B) = P(A) \qquad (3.6.2)$$

or

$$P(B|A) = P(B), \qquad (3.6.3)$$

each one of these implying the other. We obtain equations (3.6.2) and (3.6.3) from equation (3.6.1) by recalling that it is *always* true that

$$P(A \cap B) = P(A|B)P(B) = P(B|A)P(A).$$

If each of these expressions is set equal to $P(A)P(B)$, all consequences of the alternative definition follow from the original definition, provided that $P(A) \neq 0$, $P(B) \neq 0$. [If $P(A) = 0$ or $P(B) = 0$, then $P(A \cap B) = 0$ and all results are true, trivially.]

Example 1. Consider the drawing of two cards with replacement from a bridge deck (that is, the first card is replaced before the second card is drawn). The event A that the first card is a spade and the event B that the second card is a heart are clearly independent, and

$$P(A \cap B) = P(A)P(B) = \tfrac{1}{4} \cdot \tfrac{1}{4} = \tfrac{1}{16}.$$

If the cards are drawn *without* replacement, however, the events A and B are *not* independent, and

$$P(A \cap B) = P(A)P(B|A) = \tfrac{1}{4} \cdot \tfrac{13}{51} = \tfrac{13}{204}.$$

Example 2. We recall the example of Section 3.4 in which a student was selected to perform a task on Monday and another student was selected to perform a task on Wednesday. There were 10 girls and 15 boys in the class. B and G denoted, respectively, the events that a boy or girl was selected on Monday; C denoted the event that a boy was selected on Wednesday. We recall that $P(C|B) = \tfrac{14}{24}$, and $P(C|G) = \tfrac{15}{24} = \tfrac{5}{8}$. Moreover, looking at Figure 3.1 on page 55, we can see that

$$P(C) = \tfrac{3}{5} \cdot \tfrac{14}{24} + \tfrac{2}{5} \cdot \tfrac{15}{24} = \tfrac{72}{120} = \tfrac{3}{5},$$

$$P(B) = \tfrac{3}{5}, \quad \text{so that } P(C)P(B) = \tfrac{9}{25},$$

$$P(G) = \tfrac{2}{5}, \quad \text{so that } P(C)P(G) = \tfrac{6}{25}.$$

Clearly, C and B are *not* independent events, nor are C and G, by definition.

Let us now change the example so that a student, boy or girl, is eligible for duty on *both* Monday and Wednesday. Then instead of Figure 3.1 we

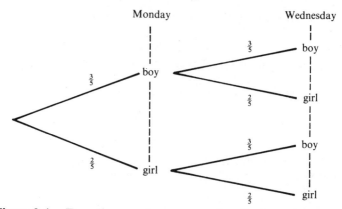

Figure 3.4. Tree diagram for an example of independent events.

have Figure 3.4. Now

$$P(C) = \tfrac{3}{5} \cdot \tfrac{3}{5} + \tfrac{2}{5} \cdot \tfrac{3}{5} = \tfrac{3}{5},$$

$$P(B) = \tfrac{3}{5},$$

$$P(G) = \tfrac{2}{5},$$

$$P(C|B) = \tfrac{3}{5},$$

$$P(C|G) = \tfrac{3}{5},$$

and events C and B are now independent; so, too, are C and G. (Apply in both cases the definition of independence of two events.)

Example 3. The events A and B in Figure 3.5, in which the points in S represent *equally likely outcomes*, are independent, because

$$n(S) = 15, \quad n(A) = 5, \quad n(B) = 6, \quad n(A \cap B) = 2,$$

$$P(A) = \tfrac{1}{3}, \quad P(B) = \tfrac{2}{5}, \quad P(A|B) = \tfrac{1}{3}, \quad P(B|A) = \tfrac{2}{5}, \quad P(A \cap B) = \tfrac{2}{15}.$$

Thus $P(A) = P(A|B)$, $P(B) = P(B|A)$, and $P(A \cap B) = P(A)P(B)$. Note that a slight change in the allocation of points in S, for example moving one

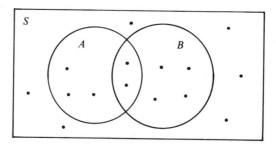

Figure 3.5. Venn diagram illustrating independent events.

point from outside A to inside A (from A' to A), would result in A and B *not* being independent.

SOME GENERAL RESULTS ON INDEPENDENCE OF EVENTS
I. For any event A in the sample space S,
   1. A and the empty, or null, event $\emptyset$ are independent.
   2. A and S are independent.

Proofs. 1. $A \cap \emptyset = \emptyset$. Hence $P(A \cap \emptyset) = P(\emptyset) = 0$. But $P(A)P(\emptyset) = P(A) \cdot 0 = 0$. Hence A and $\emptyset$ are independent, since $P(A \cap \emptyset) = P(A)P(\emptyset)$.
   2. $A \cap S = A$. Hence $P(A \cap S) = P(A)$. But $P(A)P(S) = P(A) \cdot 1 = P(A)$. Hence A and S are independent, since $P(A \cap S) = P(A)P(S)$.
II. If A and B, events in a sample space S, are independent, then so are
   1. A' and B'.
   2. A and B'.
   3. A' and B.
We suggest the reader tackle these results as exercises.

INDEPENDENCE OF THREE EVENTS A, B, AND C IN S.   The extension of the definition of independence to *any* number of events is not immediate. For this reason we now define independence of three events A, B, and C to indicate the changes necessary.
   Definition.   Events A, B, and C are independent if and only if

$$P(A \cap B) = P(A)P(B), \quad P(A \cap C) = P(A)P(C), \quad P(B \cap C) = P(B)P(C),$$

$$P(A \cap B \cap C) = P(A)P(B)P(C).$$

Thus the events must not only be independent by pairs but by triples as well.
   Example.   Suppose three balanced (fair) coins are tossed simultaneously and A, B, and C are, respectively, the events that the first coin is a head, the second coin is a head, and the third coin is a head. Then it is easy to see that

$$P(A) = P(B) = P(C) = \tfrac{1}{2},$$

$$P(A \cap B) = P(A \cap C) = P(B \cap C) = \tfrac{1}{4},$$

$$P(A \cap B \cap C) = \tfrac{1}{8}.$$

By definition, therefore, the events A, B, and C are all independent. (This is obvious from the point of view of commonsense, anyway, because the fact that a head occurs on one coin cannot affect whether a head will occur on another coin.)

INDEPENDENCE OF r EVENTS.   The r events $A_1, A_2, \ldots, A_r$ are independent if they are independent by pairs, by triples, by quadruples, and so on. That is, the events are independent if for every subset of k events,

$k = 2, 3, \ldots, r$, the probability of the intersection of the k events is equal to the product of the probabilities of the individual events. We now give two examples of the consequences of such independence.

Example 1.   Consider the repeated tossing of a balanced coin with probabilities $P(H) = P(T) = \frac{1}{2}$. If we assume that the tosses are independent (in the sense that the outcome of one toss is in no way affected by the outcome of any other toss), then, for example, the probability of getting three H's in three tosses is $P(H)P(H)P(H) = \frac{1}{2} \cdot \frac{1}{2} \cdot \frac{1}{2} = \frac{1}{8}$. If the coin is tossed 20 times and the results are 20 H's, *the probability of H on the twenty-first toss is still* $\frac{1}{2}$, because of the independence. (However, by this time, you may well begin to doubt the validity of the assumption that the coin is balanced! The probability of actually getting 20 H's in 20 tosses is extremely small if the coin is balanced, in fact, $(\frac{1}{2})^{20} = 1/1,048,576$, or less than 1 chance in 1 million!)

The idea that a long run of failures heralds a success occurs widely. In baseball games on television it is often said that, for example, "Smith is batting 0.333 and has had three 'outs' in a row so that he is now 'due'." This is, of course, nonsense. Smith's chance of making a hit is still about $\frac{1}{3}$, as it was on previous occasions. To say he is "due" implies a much higher probability.

Example 2.   Suppose that six dice are tossed simultaneously. We assume that, on each die, each of the six faces has the same chance ($\frac{1}{6}$) of turning up and that the outcome on each die is independent of the outcome on all the other dice; both of these are reasonable assumptions. Then the probability of getting all sixes is

$$P(6)P(6) \cdots P(6) = \left(\frac{1}{6}\right)^6 = \frac{1}{46,658}.$$

We also get the same answer for all fives, all fours, and so on. What is the probability that all dice show the same face? This is given by P (all sixes or all fives or ... or all ones) and, because the six events are mutually exclusive, we can add individual probabilities as follows:

$$P(\text{all dice show the same face}) = \sum_{i=1}^{6} P(\text{all dice show face } i)$$

$$= \sum_{i=1}^{6} \left(\frac{1}{6}\right)^6$$

$$= 6\left(\frac{1}{6}\right)^6.$$

$$= \frac{1}{7776}.$$

Warning. The reader should not confuse the concepts of mutually exclusive events and independent events and should reread the two definitions carefully. We often want to know whether events are mutually exclusive when we are concerned with the union of the events; we are usually interested in independence when we are concerned with the intersection of the events.

Note that, if A and B *are* mutually exclusive, then $P(A \cap B) = 0$. Since $P(A \cap B) = P(A)P(B|A) = P(B)P(A|B)$, we have $P(B|A) = P(A|B) = 0$, if $P(A) \neq 0$ and $P(B) \neq 0$, respectively. Consequently, unless one of A or B is equal to $\emptyset$ the null event, A and B are *not* independent when they are mutually exclusive. (Intuitively we see that if A and B are mutually exclusive, the outcome of B certainly depends on A, for if A occurs then B cannot occur.) If one of A or B *is* equal to $\emptyset$, then the general result I(1) given earlier in this section (p. 71) applies.

Exercises

1. If A and B are independent events and A' and B' are their respective complements, show that the following pairs of events are also independent:

    (a) A and B'.   (b) A' and B'.

2. Four cards are drawn one at a time with replacement from a standard bridge deck. Find the probability of getting (in this order) a club, a face card, a queen, and a red card.

3. If $P(A|B) = 0.6$ and $P(B|A) = 0.5$, and A and B are independent, find $P(A \cup B)$.

4. Events A and B are independent; $P(A) = 0.30$, $P(B) = 0.55$. Find $P(A \cup B)$.

5. Events A and B are independent; $P(A) = 0.6$ and $P(B) = 0.5$. Find

    (a) $P(A \cap B)$.   (b) $P(A \cup B)$.   (c) $P(A|B)$.

6. Events A and B are independent; $P(A) = 0.7$, $P(B) = 0.8$. Find

    (a) $P(A \cap B)$.   (b) $P(A \cup B)$.   (c) $P(A' \cap B')$.

7. $P(B) = \frac{1}{2}$, $P(A \cap B) = \frac{1}{3}$, and $P(A) = \frac{3}{4}$.

    (a) Find $P(A|B)$.

    (b) Find $P(A')$.

    (c) Find $P(A \cup B)$.

    (d) Is $S = A \cup B$, where S is the sample space?

    (e) Are A and B independent?

    (f) Are A and B mutually exclusive?

8. $P(A) = \frac{1}{2}$ and $P(B|A) = \frac{3}{4}$.

    (a) Find $P(A \cap B)$.

    (b) If A and B are independent, what is $P(B)$?

    (c) If A and B are independent, find $P(A \cup B)$.

    (d) Are A and B mutually exclusive?

9. Given $P(A) = 0.5$, $P(B|A) = 0.9$, find $P(A \cap B)$. If A and B are independent, what is $P(B)$? Are A and B mutually exclusive? If, instead, $P(B) = 0.6$, what are $P(A|B)$ and $P(A \cup B)$?

10. Box I contains 7 red and 4 white balls; box II contains 3 red and 6 white balls. One ball is drawn from each box. Find the probability that
    (a) one ball is red and the other is white.
    (b) both balls are red.
    (c) both balls are white.
    (d) at least one ball is white.
    (e) both balls are of the same color (red *or* white).
    (f) the ball from box I is red and the ball from box II is white.

11. A nickel and a dime are tossed. Find the probability of
    (a) 2 heads.
    (b) 1 head and 1 tail.
    (c) at least 1 head.
    (d) the nickel showing heads and the dime showing tails.
    (e) the two coins matching (that is, both heads or both tails).

12. A coin is tossed four times in succession. Find the probability of getting
    (a) *HHHT*, in that exact order.
    (b) *HTHT*, in that exact order.
    (c) *THHT*, in that exact order.
    (d) two *H*'s and two *T*'s, in any order.
    (e) four *H*'s.
    (f) no *H*'s.

13. Box I contains 1 orange, 5 apples, and 3 pears. Box II contains 2 oranges, 3 apples, and 4 pears. Two pieces of fruit are taken at random from box I and one piece from box II. Find the probability of getting
    (a) 2 oranges and 1 apple.   (c) 2 apples and 1 pear.
    (b) 3 oranges.   (d) 1 orange, 1 apple, and 1 pear.

3.7. General Addition and Multiplication Laws for k Events

In this brief section we provide the extension, to k events, of the results given earlier in Sections 3.3 to 3.5. First, however, we repeat the results for $k = 2$.

GENERAL ADDITION LAW

1. $k = 2$. If A and B are any two events,

$$P(A \cup B) = P(A) + P(B) - P(A \cap B).$$

If A and B are mutually exclusive, $A \cap B = \varnothing$, and $P(A \cap B) = 0$.

2. $k = k$. If $A_1, A_2, \ldots, A_k$ are any k events,

$$
\begin{aligned}
P(A_1 \cup A_2 \cup \cdots \cup A_k) = {}& \sum_{j=1}^{k} P(A_j) \\
& - \sum_{i=1}^{k} \sum_{j>i}^{k} P(A_i \cap A_j) \\
& + \sum_{i=1}^{k} \sum_{j>i}^{k} \sum_{l>j}^{k} P(A_i \cap A_j \cap A_l) \\
& \vdots \\
& + (-1)^{k+1} P(A_1 \cap A_2 \cap \cdots \cap A_k).
\end{aligned}
$$

In this formula the signs before the summations alternate, and each successive summation contains intersections of one more event than is contained in the intersections in the previous summation. Each summation is over all possible different arrangements. (The reader will find it helpful to write out the cases for $k = 3$ and 4.) If $A_1, A_2, \ldots, A_k$ are all mutually exclusive, then all terms except those in the first summation on the right-hand side reduce to zero.

GENERAL MULTIPLICATION LAW

1. $k = 2$. If A and B are any two events,

$$P(A \cap B) = P(A)P(B|A) = P(B)P(A|B).$$

If A and B are independent, then

$$P(B|A) = P(B) \qquad \text{and} \qquad P(A|B) = P(A).$$

2. $k = k$. If $A_1, A_2, \ldots, A_k$ are any k events,

$$
\begin{aligned}
P(A_1 \cap A_2 \cap \cdots \cap A_k) = {}& P(A_1)P(A_2|A_1)P(A_3|A_1 \cap A_2) \cdots \\
& \cdots P(A_k|A_1 \cap \cdots \cap A_{k-1}).
\end{aligned}
$$

If $A_1, A_2, \ldots, A_k$ are all independent,

$$P(A_2|A_1) = P(A_2), \quad P(A_3|A_1 \cap A_2) = P(A_3),$$
$$\ldots, P(A_k|A_1 \cap A_2 \cap \cdots \cap A_{k-1}) = P(A_k).$$

Exercises

1. Write out the general addition and multiplication laws for $k = 4$.
2. Comment on the following. There either are or are not humans on Mars, and with no evidence either way we take the probability of no humans on Mars to be $\frac{1}{2}$. Likewise, the probability of no dogs on Mars is taken to be $\frac{1}{2}$, the probability of no cats on Mars is $\frac{1}{2}$, the probability of no elephants on Mars is $\frac{1}{2}$, and so on. Thus the probability that there is none of N possible kinds of life on Mars is $(\frac{1}{2})^N$. For large N this is close to zero, so that there is almost certainly life on Mars.

CHAPTER 4

Random Variables, Distributions, and the Binomial Distribution

4.1. Idea of a Random Variable

In practical applications of the theory of probability, it is usually more convenient to work with numbers than with outcomes such as "heads," "tails," "boy," "girl," "Yankee," "Rebel." It is also mathematically easier to define functions of real numbers than functions of objects. For these reasons we now introduce the idea of a random variable.

Definition. A *random variable* is a function X that associates with each element or point s of a sample space S a real number $X(s)$, the specific value of X at the point s.

Thus a random variable is a real-valued function defined on a sample space. We recall that a probability function P is also defined on a sample space. The difference is that a probability function is defined on the set of *subsets* of the space, whereas a random variable is defined on the individual sample *points* only. The range of a probability function is the subset of real numbers between zero and one inclusive, whereas the range of a random variable is not necessarily so restrictive.

NOTATION.   The set of all sample points s such that $X(s) = x$, where x is any specified number, is written $\{X(s) = x\}$. The probability of the event $\{X(s) = x\}$, $P(\{X(s) = x\})$, will be abbreviated $P(X = x)$ or, later, $p(x)$.

Example 1.   A fair (or balanced) coin is tossed once. We can write $S \equiv \{H, T\}$, the sample space consisting of the only two possibilities head (H) and tail (T). H and T are the two points of S, so each is an "s" in the notation above. Let the two corresponding values of $X(s)$ be defined as

$$X(H) \equiv 1 \qquad X(T) \equiv 0.$$

Since, for a fair coin, head and tail are equally likely, each outcome having a probability of $\frac{1}{2}$, we can write

$$P(H) = P(X = 1) = \tfrac{1}{2},$$
$$P(T) = P(X = 0) = \tfrac{1}{2}.$$

That is, X is a random variable taking the two possible values zero and one each with a probability $\frac{1}{2}$. We illustrate this in Figure 4.1.

Example 2.   Two balanced coins are tossed simultaneously, once. We can define a sample space S consisting of the four possible outcomes and define values of X associated with these outcomes as follows:

Possible outcomes	(T, T)	(T, H)	(H, T)	(H, H)
Associated value, $X(s)$	0	1	1	2

Then, since the four possible outcomes are equally likely with probability $\frac{1}{4}$, we see that

$$P(X = 0) = \tfrac{1}{4},$$
$$P(X = 1) = \tfrac{1}{4} + \tfrac{1}{4} = \tfrac{1}{2},$$
$$P(X = 2) = \tfrac{1}{4}.$$

That is, X is a random variable taking the three possible values 0, 1, and 2 with probabilities $\frac{1}{4}$, $\frac{1}{2}$, and $\frac{1}{4}$, respectively. We illustrate this in Figure 4.2.

Figure 4.1.   Representation of the probabilities associated with a random variable X connected with a single toss of a fair coin.

Figure 4.2. Representation of the probabilities associated with a random variable X connected with a single simultaneous toss of two coins.

(In introducing probabilities at this stage we are, in fact, anticipating the specification of a probability function to be made in Section 4.3. However these simple examples of probability functions will help us to understand more easily what is involved in general. We now revert to a discussion of what has been achieved by defining a random variable X above.)

Our definition of a random variable is essentially a transformation which takes events (subsets of S) into one or more numbers (subsets of R, the set of real numbers). Inasmuch as these events arise from a random phenomenon, the values of a random variable can also be interpreted as arising from a random phenomenon. As we have already seen in the simple examples above, the values of a random variable X can be conveniently represented by subsets of real numbers on a horizontal axis. An event in S, a subset $\{s_1, s_2, \ldots, s_k\}$ of S, defines a corresponding subset $\{X(s_1), X(s_2), \ldots, X(s_k)\}$ which can itself be called an event. An interval of real numbers $[a, b]$, that is, all real numbers $X(s)$ such that $a \leq X(s) \leq b$, can constitute an event. The null set (that is, no points at all in R—not zero, for zero *can* represent an event) and the set R of all real numbers can also be events.

We can regard the random variable X as a mapping from the original sample space S to a set of real numbers R, which then becomes the new sample space. Thus if a probability function P is defined on a sample space S and a random variable X is also defined on S, the probability function can then be defined directly on the set R of values of the random variable X in such a way that a value of the random variable and the corresponding subset of points in S will have the same probability.

Suitable values of a random variable may even be specified by the problem under study. If the elements of a sample space S are real numbers, these actual numbers can be taken as the values of a random variable. Thus, if a single die is tossed, the six possible outcomes $1, 2, 3, \ldots, 6$ can be used.

Example. Suppose a hand of 13 cards is dealt from a bridge deck, and let the sample space S be the set of all hands (subsets) of size 13 from the set of 52 cards. Then each sample point s of S consists of one of the $\binom{52}{13}$ ways of allocating 13 of the 52 cards to the hand. Let X have the value $X(s) = 0, 1,$

2, 3, or 4 at each sample point s according to the number of aces in the hand, 0 for no aces, 1 for one ace, and so on. Then it is clear that X maps every sample point s of S into one of the points 0, 1, 2, 3, or 4 in R.

Exercises

1. Six slips of paper are marked a, a, a, b, b, and c, and three slips are to be drawn at random, without replacement. Describe a sample space for the experiment and define a random variable on the sample space.
2. Six cards are to be drawn at random and without replacement from a bridge deck. A sample space is the set of all possible six-card hands. Define a random variable on this sample space.
3. There are 12 girls in a mathematics class; 4 are called Peggy, 3 are called Kathy, 2 are called Mary, 1 is called Jean, 1 is called Sue, and 1 is called Nancy. Three of the girls are to be selected at random for a special assignment. Describe a sample space and define a random variable on the sample space.
4. Three dice are to be tossed simultaneously. Describe a sample space and define a random variable on the sample space.

4.2. Discrete and Continuous Random Variables

Random variables may be either discrete, continuous, or mixed.

A random variable X is said to be *discrete* if it takes only a finite or countably infinite[1] number of distinct values. The sum of the probabilities for all possible values of X is 1 in either case.

[1] There are two basic types of infinite sets:

1. One is illustrated by the set W of positive integers. To describe this set by listing the elements would be physically impossible, for we would start with 1, 2, 3, ... and no matter how long we wrote down positive integers, we would never actually finish the enumeration. Nevertheless, we say that the set of positive integers is *countably infinite* because, although we can never actually reach the end of the set, we do know how to count up all the elements in the set without missing any. Any other infinite set that can be put into one-to-one correspondence with the set of positive integers is also countably infinite.

A random variable X is said to be *continuous* if it takes a noncountably infinite number of distinct values and the probability that X takes any particular value is zero.

(A random variable that has characteristics of both discrete and continuous variables is called a mixed random variable.)

Examples. The two common types of electric clock furnish examples of discrete and continuous random variables. One type of clock has a minute hand that remains in a fixed position for 1 minute and then jumps to the next position. If we assume that the change in position is instantaneous, then the position of the minute hand is (for all practical purposes) a *discrete* random variable; the hand can occupy only 60 different specified positions. In the other type of clock, the minute hand moves continuously around the dial, and the position of the hand can lie anywhere in the interval from 0 to 60. The position in this case is a continuous random variable.

In this book we shall be concerned almost entirely with discrete random variables. (The study of continuous and mixed random variables requires either a knowledge of calculus or an equivalent symbolism.)

Examples of sets of possible values for a discrete random variable are:

1. The set of integers from 0 to n (a finite set of values).
2. The set of nonnegative integers (a countably infinite set of values).
3. The set $\{-\frac{1}{2}, 0, \frac{1}{2}\}$ (a finite set of values).

(An example of a continuous random variable is one whose set of values is the subset of real numbers between 60 and 100. This random variable could be, for example, the maximum daily temperature during July at a certain location.)

Exercise

1. Tell whether each of the following random variables is discrete, continuous, or mixed. Give reasons for your answer.
    (a) The sum of numbers showing on three tossed dice.
    (b) The number of aces in a bridge hand.
    (c) The temperature at noon in Chicago.
    (d) The weight of children in the fourth grade.
    (e) The size of men's hats in a department store.
    (f) The fraction of males in a large group of people.

---

2. There are other infinite sets that *cannot* be put into one-to-one correspondence with the set of positive integers. A set of this type is called a *noncountably infinite* set. The term noncountably infinite signifies that the elements of the set cannot be accounted for by the assignment of positive integers. It can be shown that the set R of all real numbers is such a set; so is the set of all real numbers between any two specified real numbers.

(g) The percentage of A's in the final grades of 300 students.
(h) The number of spots on a leopard.
(i) The time it takes a person to react to a stimulus.
(j) The salary of a worker in a factory.
(k) The postage fee for a package mailed in a U.S. Post Office.
(l) The pulse rate of a hospital patient.
(m) The area of a lunar sea.
(n) The volume of a raindrop.
(o) The date of the last snowfall of the season at Calumet, Michigan.
(p) The number of errors on a typewritten page.
(q) The time between phone calls at a certain switchboard.

4.3. Probability Function (and Probability Distribution) of a Discrete Random Variable

The general definition of a probability function in Section 3.3 can now be made specific for a discrete random variable. We have a sample space S consisting of simple events s and a function X which takes the specific value $X(s) = x$ at the point s of S. We denote the probability that $X(s) = x$ by $p(X(s))$, or $p(x)$ for short. The following properties must hold:

1. $0 \le p(x) \le 1$.
2. $\sum p(x) = 1$, where the summation is over all possible values of x.

Example 1. Suppose a coin is tossed and the sample space consists of H and T; that is, $S = \{H, T\}$. We can write

$$X(H) = 1,$$

$$X(T) = 0.$$

Thus, for a head, $x = 1$ and for a tail, $x = 0$. Suppose p, $0 \le p \le 1$, is the probability of a head and $1 - p$ is the probability of a tail. Then we can write $p(x)$ as

$$p(1) = p,$$

$$p(0) = 1 - p.$$

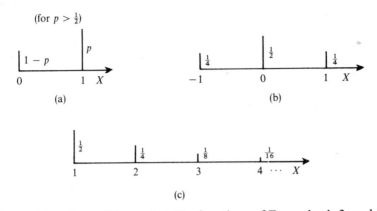

Figure 4.3.   Plots of the probability functions of Examples 1, 2, and 3.

We note that

$$\sum p(x) = p(0) + p(1) = 1 - p + p = 1.$$

This probability function is called the *Bernoulli probability function* and can be plotted as in Figure 4.3(a).

In examples we often dispense with the particular sample space and just state probabilities for the various values $X(s)$ as in the following two examples.

Example 2.   Suppose that X takes the value -1 with probability $\frac{1}{4}$, the value 0 with probability $\frac{1}{2}$, and the value 1 with probability $\frac{1}{4}$. Then we can write

$$p(-1) = \tfrac{1}{4}, \quad p(0) = \tfrac{1}{2}, \quad p(1) = \tfrac{1}{4}.$$

We see that

$$\sum p(x) = p(-1) + p(0) + p(1) = \tfrac{1}{4} + \tfrac{1}{2} + \tfrac{1}{4} = 1.$$

This probability function is plotted in Figure 4.3(b).

Example 3.   Suppose that X takes the value x with probability $1/2^x$ for $x = 1, 2, \ldots$. Then

$$p(1) = \tfrac{1}{2}, p(2) = \tfrac{1}{4}, p(3) = \tfrac{1}{8}, \ldots$$

and

$$\sum p(x) = \sum \frac{1}{2^x} = 1.$$

(Can you verify this? Do Exercise 3.) This probability function—or rather the early part of it—is plotted in Figure 4.3(c).

PROBABILITY THAT THE VALUE OF X **WILL BELONG TO A SPECIFIED SET OF VALUES.** Often we are interested in finding the probability that the value of X will belong to a specified set of values. To obtain this for a discrete random variable, we simply add the probabilities for the individual values in the specified set. (Clearly the individual values *are* mutually exclusive.)

Examples. Suppose X is a random variable with $p(x) = 1/2^x, x = 1, 2, \ldots$. Then

$$P(X < 3) = p(1) + p(2) = \tfrac{1}{2} + \tfrac{1}{4} = \tfrac{3}{4},$$

$$P(X \leq 3) = p(1) + p(2) + p(3) = \tfrac{1}{2} + \tfrac{1}{4} + \tfrac{1}{8} = \tfrac{7}{8},$$

$$P(X > 2) = p(3) + p(4) + \cdots$$

$$= 1 - p(1) - p(2) \qquad \text{[because } \sum p(x) = 1]$$

$$= 1 - \tfrac{1}{2} - \tfrac{1}{4}$$

$$= \tfrac{1}{4},$$

$$P(2 \leq X < 5) = p(2) + p(3) + p(4)$$

$$= \tfrac{1}{4} + \tfrac{1}{8} + \tfrac{1}{16}$$

$$= \tfrac{7}{16}.$$

PROBABILITY DISTRIBUTION. For a discrete random variable X, the probability function is given by $p(x)$, the probability that the random variable X assumes the value x. The set of ordered pairs $\{x, p(x)\}$ associated with a discrete random variable X is called a discrete *probability distribution*. Thus Figure 4.3 can be said to show either "plots of the probability functions" or equivalently "the probability distributions" for the examples given on pages 82 and 83.

Examples. In Example 1 we had the discrete probability distribution: $\{(0, 1 - p), (1, p)\}$; in Example 2 we had the distribution $\{(-1, \tfrac{1}{4}), (0, \tfrac{1}{2}), (1, \tfrac{1}{4})\}$; in Example 3 we had the distribution $\{(x, 1/2^x): x = 1, 2, \ldots\}$.

There are many different types of discrete distributions, but several specific ones occur so frequently in practice that it is important to study them individually. The first such distribution we shall study in Section 4.4, the *binomial distribution*, is so called because the form of $p(x)$ is related to the typical term of the expanded form of the power of a binomial expression, as we shall see later. A random variable that gives rise to the binomial distribution is called a *binomial variable*. First, however, we give some exercises on the work so far.

Exercises

1. Can one assign probabilities to individual values of the random variable whose set of possible values is the subset of all real numbers between 0 and 1? Why or why not?
2. Assign probabilities $p(x)$ to the set of x values $\{-\frac{1}{2}, 0, \frac{1}{2}\}$ so that the properties (1) $0 \leq p(x) \leq 1$, and (2) $\sum p(x) = 1$ will be satisfied. Now assign a different set of probabilities to the same set of values.
3. Verify the property $\sum p(x) = 1$ for the discrete probability distribution $\{(x, p(x) = 1/2^x) : x = 1, 2, \ldots\}$.
4. For the distribution $\{(x, p(x) = 1/2^x) : x = 1, 2, \ldots\}$, find
   (a) $P(X \leq 4)$.   (c) $P(X$ is even).
   (b) $P(X \geq 2)$.   (d) $P(X$ is a multiple of 5).
5. Suppose X takes the values $1, 2, \ldots, n$ with equal probabilities. Find, for $n > 10$,
   (a) $P(X = 6)$.   (c) $P(X > 3)$.   (e) $P(X < n - 5)$.
   (b) $P(X = k), k < n$.   (d) $P(X \geq 3)$.
6. A random variable takes the values 2, 3, 4, 7, and 8 with probabilities $\frac{3}{25}$, $\frac{6}{25}$, $\frac{5}{25}$, $\frac{7}{25}$, and $\frac{4}{25}$, respectively. Find the probability that the variable will be
   (a) an odd number.   (b) less than 7.   (c) a power of 2.
7. Suppose X takes the value 1 when the minute hand of a clock is between 12 and 1, the value 2 when the minute hand is between 1 and 2, and so on. Assign equal probabilities to each value X can take and find
   (a) $P(X = 12)$.   (c) $P(X > 1)$.
   (b) $P(X < 6)$.   (d) $P(X = 0)$.
8. From a box containing chips numbered 1 to N, two chips are withdrawn without replacement. Assume equal probabilities for all possible outcomes. Let the random variable X be "the absolute value of the difference of the numbers on the two chips." Specify the possible values for x (which X can take), and write a formula for $p(x)$.

4.4. Binomial Distribution

There are many possible discrete probability distributions; we are now going to discuss one of them. The reason we select this particular distribution to begin with is that it occurs frequently in practical probability problems. It arises, in fact, in all situations where the requirements below are satisfied. As we shall see in the exercises, a number of other important distributions also have connections with the binomial distribution.

REQUIREMENTS

1. A random phenomenon, and a sample space with only two distinct elements, or labels (see page 42). One of these labels will be called *success* and the other will be called *failure*.

2. Independent outcomes. The outcome of a particular occurrence, or *trial*, of the random phenomenon is completely independent of, and not affected by, the outcome of any other trial.

3. Constant probability of success in a single trial. The probability p that the outcome of an occurrence of the phenomenon is a success is constant. (Thus the probability $q = 1 - p$ that it is a failure is also constant.)

Then, if requirements 1, 2, and 3 all hold, and n trials are made, where n is fixed, the number of successes is a *binomial variable* (so is the number of failures) and has a *binomial probability distribution*.

Example. A fair die is thrown five times ($n = 5$ trials). A success is recorded if the result is a six, a failure if the result is not a six. Since all six faces are equally likely to occur, $p = \frac{1}{6}$, $q = \frac{5}{6}$. The number of sixes, X, is a binomial variable, because requirements 1, 2, and 3 are all satisfied. [(1) A six is a success; any other result, a failure; (2) the outcome on each toss is independent of the outcomes on all other tosses; (3) $p = \frac{1}{6}$ is constant on each toss.]

Now X can take the values $x = 0, 1, 2, 3, 4,$ or 5. We shall want $p(x)$ for all these x's. Let us first consider $p(3)$ for purposes of example.

In how many ways can we get three sixes (successes, S) and two "not-sixes" (failures, F) in five throws? the answer is

$$\binom{5}{3} = \binom{5}{2} = 10,$$

since[2] the number of outcomes belonging to the event "exactly three sixes in five tosses" is the same as the number of permutations of five objects where there are three of one kind and two of another kind, namely $5!/3!2! = 10$. One of these 10 results is the sequence $SSSFF$. What is the

---

[2] Note that, when there are only two different kinds of objects, the number of permutations of n objects when there are n_1 of one kind and n_2 of a second kind is the same as the number of combinations of n_1 (or n_2) objects taken from a set of n objects, namely $\binom{n}{n_1} = \binom{n}{n_2}$. We will use the combination notation in the binomial distribution.

probability of obtaining this *particular* sequence? Since the trials are independent,

$$P(SSSFF) = P(S) \cdot P(S) \cdot P(S) \cdot P(F) \cdot P(F)$$
$$= p \cdot p \cdot p \cdot (1 - p) \cdot (1 - p)$$
$$= p^3 (1 - p)^2$$
$$= (\tfrac{1}{6})^3 (\tfrac{5}{6})^2.$$

Now the $\binom{5}{3} = 10$ individual sequences of three successes and two failures are mutually exclusive—only one particular sequence of three successes and two failures can occur in a particular set of five trials—and each has probability $(\tfrac{1}{6})^3(\tfrac{5}{6})^2$. Thus the probability of the required event is

$$p(3) = \binom{5}{3}\left(\frac{1}{6}\right)^3\left(\frac{5}{6}\right)^2.$$

[Using the tree diagram, Figure 4.4, we merely add together the probabilities for all those points with three S's and two F's; there are $\binom{5}{3} = 10$ of these points, each with probability $(\tfrac{1}{6})^3(\tfrac{5}{6})^2$.]

The full set of $p(x)$ is as follows:

$$p(0) = \binom{5}{0}\left(\frac{1}{6}\right)^0\left(\frac{5}{6}\right)^5 = \frac{3125}{7776},$$

$$p(1) = \binom{5}{1}\left(\frac{1}{6}\right)^1\left(\frac{5}{6}\right)^4 = \frac{3125}{7776},$$

$$p(2) = \binom{5}{2}\left(\frac{1}{6}\right)^2\left(\frac{5}{6}\right)^3 = \frac{1250}{7776},$$

$$p(3) = \binom{5}{3}\left(\frac{1}{6}\right)^3\left(\frac{5}{6}\right)^2 = \frac{250}{7776},$$

$$p(4) = \binom{5}{4}\left(\frac{1}{6}\right)^4\left(\frac{5}{6}\right)^1 = \frac{25}{7776},$$

$$p(5) = \binom{5}{5}\left(\frac{1}{6}\right)^5\left(\frac{5}{6}\right)^0 = \frac{1}{7776}.$$

Note that, since $3125 + 3125 + 1250 + 250 + 25 + 1 = 7776$, it follows that

$$\sum_{x=0}^{5} p(x) = 1,$$

as required.

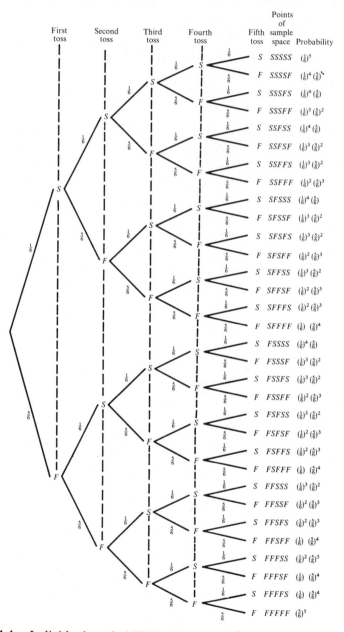

Figure 4.4. Individual probabilities of all the $2^5 = 32$ possible ways of obtaining sixes in five throws of a die.

THE BINOMIAL THEOREM. For the general binomial distribution we shall need to know the following general result, known as the binomial theorem.

$$(y + z)^n = y^n + ny^{n-1}z + \tfrac{1}{2}n(n - 1)y^{n-2}z^2 + \cdots + nyz^{n-1} + z^n$$

$$= \binom{n}{0}y^n + \binom{n}{1}y^{n-1}z + \binom{n}{2}y^{n-2}z^2 + \cdots + \binom{n}{n-1}yz^{n-1} + \binom{n}{n}z^n$$

$$= \sum_{x=0}^{n} \binom{n}{x}y^x z^{n-x}.$$

We omit a proof of this result and mention only that it is easily proved by induction, by taking the following steps:

1. Write out the result above with $n - 1$ instead of n everywhere; accept this as true for the moment.

2. Multiply both sides of the equation in step 1 by $(y + z)$. The left-hand side is $(y + z)^n$, and the right-hand-side terms, which result from the product, can be collected and rearranged to give the binomial theorem above, by using the fact that $\binom{n-1}{r-1} + \binom{n-1}{r} = \binom{n}{r}$, as is easily shown algebraically.

Thus, *if* the result is true for $n - 1$, it is true for n also.

3. Now for $n = 1$, the binomial theorem says that

$$y + z = y + z$$

and is *obviously* true. Hence by step 2 it is true for $n = 2, 3, \ldots$, that is, for all values of n.

APPLICATIONS. Let $y = z = 1$. Then by the binomial theorem

$$\sum_{x=0}^{n} \binom{n}{x}1^x 1^{n-x} = (1 + 1)^n$$

or

$$\binom{n}{0} + \binom{n}{1} + \binom{n}{2} + \cdots + \binom{n}{n-1} + \binom{n}{n} = 2^n.$$

(We mentioned this result in Section 1.5, page 20, but used a different argument there to obtain it.) A similar application with $y = 1$, $z = -1$ enables us to prove that

$$\binom{n}{0} - \binom{n}{1} + \binom{n}{2} - \binom{n}{3} + \cdots + (-1)^n\binom{n}{n} = 0.$$

The reader can confirm this result for specific examples by writing down the left-hand side for a few selected low values of n. Binomial coefficients

$$
\begin{array}{ccccccccccccc}
& & & & & & & 1 & & & & & \\
n = 1 \rightarrow & & & & & & 1 & & 1 & & & & \\
n = 2 \rightarrow & & & & & 1 & & 2 & & 1 & & & \\
n = 3 \rightarrow & & & & 1 & & 3 & & 3 & & 1 & & \\
n = 4 \rightarrow & & & 1 & & 4 & & 6 & & 4 & & 1 & \\
n = 5 \rightarrow & & 1 & & 5 & & 10 & & 10 & & 5 & & 1 \\
n = 6 \rightarrow & 1 & & 6 & & 15 & & 20 & & 15 & & 6 & & 1 \\
\end{array}
$$

Figure 4.5. Pascal's triangle for obtaining binomial coefficients.

can be quickly obtained for small n by using Pascal's triangle (Figure 4.5), where each row is obtained from the one above by summing the two figures above it. (Do Exercise 28, page 102.)

THE GENERAL FORMULA FOR THE BINOMIAL DISTRIBUTION.

The general formula for the binomial distribution is derived for n independent trials (where X is the number of successes) when the probability of success at each trial is p and the probability of failure at each trial is $1 - p$. The probability of exactly x successes in the n trials, that is, the probability that $X = x$, is given by

$$
p(x) = \binom{n}{x} p^x (1 - p)^{n-x} \qquad x = 0, 1, 2, \ldots, n.
$$

The argument for getting this consists of noting that there are $\binom{n}{x}$ mutually exclusive outcomes belonging to the event "x successes in n trials," each of which has probability $p^x (1 - p)^{n-x}$ (since each of the x successes has probability p, and each of the $n - x$ failures has probability $q = 1 - p$).

Now the sum of the probabilities over all possible values of the random variable must be one; that is,

$$
\sum_{x=0}^{n} \binom{n}{x} p^x (1 - p)^{n-x} = 1.
$$

We can see this by appealing to the binomial theorem. We obtain, setting $y = p, z = 1 - p$,

$$
\sum_{x=0}^{n} \binom{n}{x} p^x (1 - p)^{n-x} = \{p + (1 - p)\}^n = 1^n = 1.
$$

It is because of its close connection with the binomial theorem that this distribution is called the binomial distribution. The coefficients $\binom{n}{x}$ are called the *binomial coefficients* for the same reason. The letters n and p are called the *parameters* of the distribution.

Example.   A card is drawn at random from an ordinary bridge deck; a success is recorded if the card is a spade and a failure is recorded otherwise. This constitutes one trial. The drawn card is returned to the deck and a second trial is made. Altogether n trials are made and the number of successes in these trials, X, is noted.

Clearly X is a binomial variable, because (1) a designation of successes and failures exists, (2) trials are independent, and (3) there is a constant probability of success $p = \frac{1}{4}$.

Suppose we ask: If $n = 5$ trials are made, what is the probability of getting $x = 4$ spades? Using the binomial distribution result, the answer is

$$p(4) = \binom{5}{4}\left(\frac{1}{4}\right)^4\left(\frac{3}{4}\right)^1 = \frac{15}{4^5} = \frac{15}{1024}.$$

What is the probability of getting at least 4 spades? In other words, what is $P(X = 4 \text{ or } X = 5)$? Clearly the events $\{X = 4\}$ and $\{X = 5\}$ are mutually exclusive; thus, as in the examples in Section 4.3, we add the individual probabilities to give

$$P(X = 4 \quad \text{or} \quad X = 5) = p(4) + p(5)$$

$$= \binom{5}{4}\left(\frac{1}{4}\right)^4\left(\frac{3}{4}\right)^1 + \binom{5}{5}\left(\frac{1}{4}\right)^5\left(\frac{3}{4}\right)^0$$

$$= \frac{15}{1024} + \frac{1}{1024}$$

$$= \frac{1}{64}.$$

CUMULATIVE BINOMIAL PROBABILITIES.   The last part of the example contained an important idea, that of the probability of getting "r or more than r successes" for $r = 0, 1, 2, \ldots, n$. We shall write this probability in general as

$$P(X \geq r) = B_r = B_r(n, p) = \sum_{x=r}^{n} p(x) = \sum_{x=r}^{n} \binom{n}{x}p^x(1 - p)^{n - x}.$$

Computing both the $p(x)$'s and the B_r's can be quite tedious. Fortunately excellent tables of both $p(x)$ and B_r are available. A very useful source is *Handbook of Tables for Probability and Statistics*, W. H. Beyer (ed.), Chemical Rubber Co., Cleveland, 2nd ed., 1968. (There are two versions of this, un-abridged and abridged. We particularly recommend the unabridged version

Table 4.1. Cumulative Binomial Probabilities $B_r(n, p)$ (Entries omitted are less than .0005)

						p					
n	r	.01	.05	.10	.16[a]	.20	.25	.30	.33	.40	.50
1	1	.010	.050	.100	.167	.200	.250	.300	.333	.400	.500
2	1	.020	.098	.190	.306	.360	.438	.510	.556	.640	.750
	2		.002	.010	.028	.040	.063	.090	.111	.160	.250
3	1	.030	.143	.271	.421	.488	.578	.657	.704	.784	.875
	2		.007	.028	.074	.104	.156	.216	.259	.352	.500
	3			.001	.005	.008	.016	.027	.037	.064	.125
4	1	.039	.185	.344	.518	.590	.684	.760	.802	.870	.938
	2	.001	.014	.052	.132	.181	.262	.348	.407	.525	.688
	3			.004	.016	.027	.051	.084	.111	.179	.312
	4				.001	.002	.004	.008	.012	.026	.062
5	1	.049	.226	.410	.598	.672	.763	.832	.868	.922	.969
	2	.001	.023	.081	.196	.263	.367	.472	.539	.663	.812
	3		.001	.009	.035	.058	.104	.163	.210	.317	.500
	4				.003	.007	.016	.031	.045	.087	.188
	5						.001	.002	.004	.010	.031
6	1	.059	.265	.469	.665	.738	.822	.882	.912	.953	.984
	2	.001	.033	.114	.263	.345	.466	.580	.649	.767	.891
	3		.002	.016	.062	.099	.169	.256	.320	.456	.656
	4			.001	.009	.017	.038	.070	.100	.179	.344
	5				.001	.002	.005	.011	.018	.041	.109
	6							.001	.001	.004	.016
7	1	.068	.302	.522	.721	.790	.867	.918	.941	.972	.992
	2	.002	.044	.150	.330	.423	.555	.671	.737	.841	.938
	3		.004	.026	.096	.148	.244	.353	.429	.580	.773
	4			.003	.018	.033	.071	.126	.173	.290	.500
	5				.002	.005	.013	.029	.045	.096	.227
	6						.001	.004	.007	.019	.062
	7									.002	.008
8	1	.077	.337	.570	.767	.832	.900	.942	.961	.983	.996
	2	.003	.057	.187	.395	.497	.633	.745	.805	.894	.965
	3		.006	.038	.135	.203	.322	.448	.532	.685	.855
	4			.005	.031	.056	.114	.194	.259	.406	.637
	5				.005	.010	.027	.058	.088	.174	.363
	6					.001	.004	.011	.020	.050	.145

Table 4.1.—*continued*

n	r	.01	.05	.10	.16[a]	.20	.25	.30	.33	.40	.50
	7							.001	.003	.009	.035
	8									.001	.004
9	1	.086	.370	.613	.806	.866	.925	.960	.974	.990	.998
	2	.003	.071	.225	.457	.564	.700	.804	.857	.929	.980
	3		.008	.053	.178	.262	.399	.537	.623	.768	.910
	4		.001	.008	.048	.086	.166	.270	.350	.517	.746
	5			.001	.009	.020	.049	.099	.145	.267	.500
	6				.001	.003	.010	.025	.042	.099	.254
	7						.001	.004	.008	.025	.090
	8								.001	.004	.020
	9										.002
10	1	.096	.401	.651	.838	.893	.944	.972	.983	.994	.999
	2	.004	.086	.264	.515	.624	.756	.851	.896	.954	.989
	3		.012	.070	.225	.322	.474	.617	.701	.833	.945
	4		.001	.013	.070	.121	.224	.350	.441	.618	.828
	5			.002	.015	.033	.078	.150	.213	.367	.623
	6				.002	.006	.020	.047	.077	.166	.377
	7					.001	.004	.011	.020	.055	.172
	8							.002	.003	.012	.055
	9									.002	.011
	10										.001
15	1	.140	.537	.794	.935	.965	.987	.995	.998	1.000	1.000
	2	.010	.171	.451	.740	.833	.920	.965	.981	.995	1.000
	3		.036	.184	.468	.602	.764	.873	.921	.973	.996
	4		.006	.056	.232	.352	.539	.703	.791	.910	.982
	5		.001	.013	.090	.164	.314	.485	.596	.783	.941
	6			.002	.027	.061	.148	.278	.382	.597	.849
	7				.007	.018	.057	.131	.203	.390	.696
	8				.001	.004	.017	.050	.088	.213	.500
	9					.001	.004	.015	.031	.095	.304
	10						.001	.004	.009	.034	.151
	11							.001	.002	.009	.059
	12									.002	.018
	13										.004
	14										.001
	15										

Table 4.1.—*continued*

					p						
n	r	.01	.05	.10	$.16^a$	.20	.25	.30	.33	.40	.50
20	1	.182	.642	.878	.974	.988	.997	.999	1.000	1.000	1.000
	2	.017	.264	.608	.870	.931	.976	.992	.997	.999	1.000
	3	.001	.075	.323	.671	.794	.909	.965	.982	.996	1.000
	4		.016	.133	.433	.589	.775	.893	.940	.984	.999
	5		.003	.043	.231	.370	.585	.762	.848	.949	.994
	6			.011	.102	.196	.383	.584	.703	.874	.979
	7			.002	.037	.087	.214	.392	.521	.750	.942
	8				.011	.032	.102	.228	.339	.584	.868
	9				.003	.010	.041	.113	.191	.404	.748
	10				.001	.003	.014	.048	.092	.245	.588
	11					.001	.004	.017	.038	.128	.412
	12						.001	.005	.013	.057	.252
	13							.001	.004	.021	.132
	14								.001	.006	.058
	15									.002	.021
	16										.006
	17										.001
	18										
	19										
	20										

$^a .1\dot{6} = \frac{1}{6}; .\dot{3}\dot{3} = \frac{1}{3}.$

because it contains all the usual statistical tables and a large amount of other useful statistical material as well.) The *Handbook* contains individual binomial terms $p(x)$ and cumulative terms for values of $n = 1(1)20$, $x = 1(1)n$, $p = 0.05(0.05)0.50$, to four decimal places. [*Note:* $n = 1(1)20$ means "values of n from 1 to 20 in intervals of 1" (the parenthetical figure).]

A short table of cumulative binomial probabilities B_r is given as Table 4.1. Note that, if individual binomial terms are required, they can be obtained by taking the difference between successive cumulative terms since, for any values of n and p,

$$p(x) = B_x - B_{x+1}.$$

BINOMIAL VARIABLE AS THE SUM OF INDEPENDENT BER-NOULLI VARIABLES. We may also think of the binomial distribution as the distribution of a sum of independent Bernoulli random variables. We recall from Section 4.3, Example 1, that the Bernoulli probability function has $p(0) = 1 - p$ and $p(1) = p$ for the two possible X values $x = 0$ and $x = 1$. Suppose now that a random phenomenon which follows a Bernoulli probability law is to be observed n times. What is the probability that the outcome 1 will occur x times? Since the n observations are independent, the probability is $p^x(1 - p)^{n-x}$ for every possible arrangement of x ones and $(n - x)$ zeros, and so the overall probability of x ones is $\binom{n}{x}p^x(1 - p)^{n-x}$, because there are $\binom{n}{x}$ possible arrangements.

Example 1. A wheel used in gambling can stop in 18 equally likely positions numbered 1 to 18. A customer places bets on all the positions divisible by 3. What is the probability that he will win exactly 3 times in 8 attempts?

Since the 18 positions are equally likely, the probability that the wheel will stop at any one position is $\frac{1}{18}$. Since there are 6 numbers exactly divisible by 3, between 1 and 18 inclusive, the probability that the customer will win on any one attempt is $\frac{1}{3}$, and the probability that he will lose is $\frac{2}{3}$. Thus

$$p(3) = \binom{8}{3}\left(\frac{1}{3}\right)^3\left(\frac{2}{3}\right)^5 = \frac{1792}{6561} = 0.273.$$

What is the probability that the customer will win 3 *or more* times in 8 attempts? Since $p = \frac{1}{3}$, we can look this up in Table 4.1, but we shall also evaluate it here as a check:

$$\sum_{x=3}^{8} p(x) = \sum_{x=0}^{8} p(x) - \sum_{x=0}^{2} p(x)$$

$$= 1 - \sum_{x=0}^{2} p(x)$$

$$= 1 - p(0) - p(1) - p(2).$$

[Note that by using the fact that the total probability equals one, we need work out only three $p(x)$'s rather than six.] Now

$$p(0) = \binom{8}{0}\left(\frac{1}{3}\right)^0\left(\frac{2}{3}\right)^8 = \frac{256}{6561},$$

$$p(1) = \binom{8}{1}\left(\frac{1}{3}\right)^1\left(\frac{2}{3}\right)^7 = \frac{1024}{6561},$$

$$p(2) = \binom{8}{2}\left(\frac{1}{3}\right)^2\left(\frac{2}{3}\right)^6 = \frac{1792}{6561}.$$

Thus

$$\sum_{x=3}^{8} p(x) = 1 - \frac{256 + 1024 + 1792}{6561} = \frac{3489}{6561} = 0.532.$$

In other words, there is a 53.2% chance that the customer will win 3 or more times in 8 plays.

Example 2. Another person selects a card at random from a bridge deck, notes its value, and returns the card to the deck. Three more independent selections are made in the same manner. You are told that at least one of the four values recorded is an ace. What odds would you take that exactly two of the four values are aces?

Let A and B be the events that at least one card is an ace and exactly two cards are aces, respectively. We thus need the conditional probability $P(B|A) = P(B \cap A)/P(A)$. Since the event B is a subset of the event A, $B \cap A = B$; it follows that

$$P(B|A) = P(B)/P(A) = \binom{4}{2}\left(\frac{1}{13}\right)^2\left(\frac{12}{13}\right)^2 \bigg/ \sum_{x=1}^{4} \binom{4}{x}\left(\frac{1}{13}\right)^x\left(\frac{12}{13}\right)^{4-x}$$

The numerator of this fraction has value 864/28,561, and the denominator is equal to

$$1 - \binom{4}{0}\left(\frac{1}{13}\right)^0\left(\frac{12}{13}\right)^4 = 1 - \frac{20,736}{28,561}$$

$$= \frac{7,825}{28,561}$$

because

$$\sum_{x=0}^{4} \binom{4}{x}\left(\frac{1}{13}\right)^x\left(\frac{12}{13}\right)^{4-x} = 1.$$

Thus

$$P(B|A) = \frac{864}{7825} = 0.110.$$

That is, there is an 11% chance that there are exactly two aces, given that there is at least one ace. Thus you should take odds of at least 89 to 11, that is, at least about 8 to 1.

Exercises

1. Ten digits are selected at random and with replacement from the digits 0 to 9 inclusive. Find the probability of getting nine zeros.

2. If a basketball player makes 80% of the free throws he attempts, find the probability that he will make 8 of 10 in a specific game.

3. The probability that a new automobile has a windshield that does not leak is $\frac{4}{5}$. Find the probability that of six new automobiles, exactly five will have windshields that leak.

4. The probability of success in a certain binomial distribution is $\frac{1}{4}$. Find the probability of three or more successes in five trials.

5. A balanced coin is tossed four times. Find the probability of obtaining
   (a) exactly two heads.
   (b) at least two heads.
   (c) at most two heads.

6. A balanced coin is tossed six times. Find the probability of obtaining
   (a) exactly four heads.
   (b) at least four heads.
   (c) at most four heads.

7. Seven cards are drawn with replacement from an ordinary bridge deck. Find the probability that
   (a) exactly five are spades.
   (b) exactly one is an ace.
   (c) exactly four are face cards.
   (d) exactly six are black.

8. The probability that the performance of a radio tube will drop to an unsatisfactory level within 10 months is $\frac{3}{5}$. What is the probability that a six-tube radio will have exactly three unsatisfactory tubes by the time it is 10 months old? (Assume that the performance of each tube is independent of performances of all other tubes.)

9. A tetrahedron (a four-sided die) has two faces marked with a one, one face marked with a two, and one face marked with a zero. Let the face flat on the floor denote the counted score. Find the probability of getting
   (a) four twos in six tosses.   (b) four ones in six tosses.

10. A coin is biased 2 to 1 in favor of heads. Find the probability of getting
    (a) five heads in eight tosses.
    (b) eight heads in eight tosses.
    (c) at most seven heads in eight tosses.

11. George, John, and David are the only three men in a business office. One is selected each day by a random process to work an extra hour that afternoon. What is the probability that George will work
    (a) five days in a row?
    (b) three days out of five?
    (c) none of the five days?

12. A coin is biased so that the probability of heads is 0.6. It is tossed seven times. Find the probability of getting

(a) exactly five heads.

(b) at least five heads.

(c) at most five heads.

13. A student taking a true-false quiz tosses a coin to obtain his answers, marking true for heads and false for tails. Unknown to him, the coin is biased so that heads comes up $\frac{2}{3}$ of the time. If there are eight statements on the quiz and all are true, find the probability that he will get

(a) eight correct.

(b) four correct.

(c) at least six correct.

(d) the first four correct and the second four incorrect.

14. To construct a true-false quiz a professor tosses a balanced coin. If it is heads he composes a true statement, and if it is tails he composes a false statement. He does this ten times. A student who hasn't prepared for the quiz tosses a balanced coin to determine his answers for each statement. He answers true for heads and false for tails. What is the probability that the exam will have

(a) ten true statements?

(b) ten false statements?

(c) the first five true and the rest false?

What is the probability that the student will get

(d) ten correct answers?

(e) five correct and five incorrect?

(f) at least eight correct?

15. Two evenly matched teams play a series of seven games. What is the probability that a particular team will win

(a) exactly four games?

(b) all seven games?

(c) every other game beginning with the first?

(d) every other game beginning with the second?

16. **The Geometric Distribution.** If the conditions for independent binomial trials are satisfied but, instead of being interested in the number of successes in n trials as in the binomial distribution, we take the random variable X to be the number of *trials required to obtain the first success,* then we obtain what is called the *geometric distribution.* The possible values of X are $x = 1, 2, \ldots$. The probability function for this discrete random variable is

$$p(x) = (1 - p)^{x-1}p,$$

because, for the first success to appear on the xth trial with probability p, we must first have $x - 1$ failures, each with probability $1 - p$.

(a) Find the probability that the first head will appear on the fifth toss of a balanced coin.

(b) Find the probability that the first six will appear on the fourth toss of a die.

(c) Find the probability that the first six will appear on or before the fourth toss of a die.

(d) Show that $\sum_x p(x) = \sum (1 - p)^{x-1} p = 1$.

17. **The Pascal Distribution.** If the conditions for independent binomial trials are satisfied but, instead of being interested in the number of trials to the occurrence of the first success, as in the geometric distribution, or in the number of successes in n trials, as in the binomial distribution, we take the random variable X to be the number of trials required to obtain r successes, then we obtain what is called the *Pascal distribution*. The possible values of X are $x = r, r + 1, r + 2, \ldots$. The probability function for this discrete random variable is

$$p(x) = \binom{x - 1}{r - 1} p^r (1 - p)^{x-r},$$

for if the rth success appears on the xth trial with probability p, then there must be $r - 1$ successes in the first $x - 1$ trials. The probability function, then, is the product of the binomial distribution for $r - 1$ successes in $x - 1$ trials: $\binom{x - 1}{r - 1} p^{r-1}(1 - p)^{x-r}$, and the probability of success on the xth trial, p. Note that, when $r = 1$, we obtain the geometric distribution (see Exercise 16) as a special case.

(a) Find the probability that the fifth head will appear on the tenth toss of a balanced coin.

(b) Find the probability that the third six will appear on the tenth toss of a die.

(c) Find the probability that the second three will appear on or before the fourth toss of a die.

(d) Show that $\sum_x p(x) = \sum_x \binom{x - 1}{r - 1} p^r (1 - p)^{x-r} = 1$.

18. **The Negative Binomial Distribution.** If the conditions for independent binomial trials are satisfied, but we are not interested in

(i) the number of trials to obtain the rth success (the Pascal distribution—see Exercise 17)

(ii) the number of trials needed for the occurrence of the first success (the geometric distribution—see Exercise 16)

    (iii) the number of successes in n trials (the binomial distribution— see the text above)

but instead in

    (iv) the random variable X = the number of failures before the rth success,

then we want what is called the *negative binomial distribution*. The possible values of X are $x = 0, 1, 2, \ldots$. The probability function for this discrete random variable is

$$p(x) = \binom{r + x - 1}{x} p^r (1 - p)^x,$$

for in order to obtain r successes with x failures we must have $r + x$ trials; the last trial must be a success and in the first $r + x - 1$ trials there must appear x failures and $r - 1$ successes. [The reader should now be able to obtain $p(x)$ by an argument similar to that in Exercise 17.]

    (a) Find the probability that there will be four tails before getting the second head when tossing a balanced coin.

    (b) Find the probability that there will be five other results before getting the second six when tossing a die.

    (c) Show that $\sum_{x} \binom{r + x - 1}{x} p^r (1 - p)^x = 1$.

COMMENTS ON THE GEOMETRIC, PASCAL, AND NEGATIVE BINOMIAL DISTRIBUTIONS

1. If, in the Pascal distribution case we denote the number of failures $x - r$ by y, then $y = x - r$ or $x = y + r$. Substitution for x into the Pascal probability function gives the negative binomial probability function with y as the variable concerned, using the fact that

$$\binom{r + y - 1}{r - 1} = \binom{r + y - 1}{y}.$$

2. When $r = 1$, the negative binomial distribution reduces to the geometric distribution.

3. The negative binomial bears this name because we can write

$$\binom{r + x - 1}{x} = \frac{(r + x - 1)!}{x!(r - 1)!}$$

$$= \frac{(r + x - 1)(r + x - 2)\cdots(r + 1)r}{x!}$$

$$= (-1)^x \frac{(-r)(-r - 1)(-r - 2)\cdots(-r - x + 2)(-r - x + 1)}{x!}$$

$$= (-1)^x \binom{-r}{x},$$

where $\begin{pmatrix} -r \\ x \end{pmatrix}$ is a "negative binomial" coefficient and has the definition implied above. Thus we can write

$$\begin{pmatrix} r + x - 1 \\ x \end{pmatrix} p^r (1 - p)^x = \begin{pmatrix} -r \\ x \end{pmatrix} p^r (p - 1)^x,$$

where we have changed the sign in the last factor to accommodate the $(-1)^x$. This last expression is the "true" negative binomial form.

4. Negative binomial (and therefore Pascal and geometric) probabilities can be obtained via binomial probabilities. See Bernard Harris, *Theory of Probability*, Addison-Wesley, Reading, Mass., 1966, pp. 64–65.

Exercises (continued)

19. Find the probability of getting four heads in six tosses of a balanced coin given that there were exactly two heads in the first three tosses.
20. Find the probability of getting exactly four H's when tossing five balanced coins if at least one is known to be an H.
21. A man is playing with two dice. What is the probability that, in five throws of the two dice, he will get
    (a) three sevens?
    (b) at least three sevens?
    (c) three *consecutive* sevens?
    [*Note*: In (c) he may get *more* than three sevens.]
22. What is the probability p that in a bridge hand of 13 cards there is no ace? What is the probability that a person who plays $n = 5$ hands of bridge never receives an ace?
23. Show that, for binomial probabilities,

$$\begin{pmatrix} n \\ x \end{pmatrix} p^x (1 - p)^{n-x} = \begin{pmatrix} n \\ n - x \end{pmatrix} (1 - p)^{n-x} p^x,$$

and explain in words what this means.
24. Show that, for given values of $n, p, q = 1 - p$, and x, the binomial probability $p(x)$ satisfies the following *recursion formula*:

$$p(x + 1) = \frac{n - x}{x + 1} \cdot \frac{p}{q} \cdot p(x).$$

25. Ten independent Bernoulli trials are made; the probability of success on each trial is p.
    (a) If $p = 0.2$, what is the probability that 4 or more successes will occur? (Use Table 4.1.)
    (b) If $p = 0.6$, what is the probability that 4 or fewer successes will occur? (Use Table 4.1, with $n = 10$, "p" $= 1 - p = 0.4$, and

$x = 6$; that is, find the probability that 6 or more *failures*, each with probability $1 - p = 0.4$, occur.)

(c) For what value of p, approximately, does the probability that 3 or more successes occur equal 0.5? (Interpolate in the table for $n = 10$, $x = 3$, and so find a value of p for which the entry would be 0.5 if it were listed.) Use your estimated value of p to find the probability that 3 or more successes occur. (You will need log tables or more detailed tables of cumulative binomial probabilities; omit this part if you do not have either of these.)

26. You are told that n independent Bernoulli trials gave rise to exactly m successes. Find, for any particular trial, the probability of a success, *conditional on this information*. (*Hint*: The answer is m/n.)

27. The popular game of "craps" is played as follows. The "shooter" rolls two fair dice repeatedly until he wins or loses. He

(1) WINS if (a) his first throw is 7 or 11, or

          (b) his first throw is 4, 5, 6, 8, 9, or 10 *and*, in a subsequent throw, the value of his first throw (his "point") is repeated before a 7 occurs.

(2) LOSES if (a) his first throw is 2, 3, or 12, or

          (b) his first throw is 4, 5, 6, 8, 9, or 10 *and*, in a subsequent throw, a 7 occurs before the value of his first throw is repeated.

(a) Show that the probability of the shooter's winning on the first throw is $\frac{8}{36}$.

(b) Show that the probability of the shooter's subsequently winning, given that his first throw is 4, is $\frac{1}{3}$.

(c) Repeat (b) with the figures 5, $\frac{2}{5}$.

(d) Repeat (c) with the figures 6, $\frac{5}{11}$.

(e) Show that the overall probability that the shooter will win is $\frac{244}{495} \doteq 0.493$, so that the odds are slightly against him.

28. Write down the $n = 7, 8, 9$, and 10 rows of Pascal's triangle.

(For a fascinating article, "The Multiple Charms of Pascal's Triangle," by Martin Gardner, see *Scientific American*, December 1966, pages 128–131.)

CHAPTER 5

Distribution Function, Expected Value, Mean, and Variance

5.1. Distribution Function

The function p whose value $p(x)$ gives the probability that a discrete random variable X will assume the value x is called the probability function of the random variable. Suppose, however, that for a given random variable X we would like to know the probability that X will assume a value *less than or equal to* x. The function that gives this probability is called the distribution function. It is defined in terms of the probability function p. If the distribution function itself is given, we can obtain the probability function from it. (Examples are given below.) First we define the distribution function.

Definition. The *distribution function*[1] of a random variable X is the function F whose value $F(x)$ gives the probability that the random variable

[1] Note the difference between the *distribution* (which consists of all the values that the variable X can take and the associated probabilities) and the *distribution function* (which is the specific function of x defined here).

103

X will assume a value less than or equal to x. We write

$$F(x) = P(X \leq x)$$
$$= \sum_{t \leq x} p(t)$$

for a discrete random variable. [We have used the symbol t to represent possible values of X to avoid ambiguity that might arise from using x for both the possible values of X in general and the specific values of X in the expression $P(X \leq x)$.]

Example 1. If $p(1) = \frac{1}{6}$, $p(2) = \frac{1}{3}$, and $p(3) = \frac{1}{2}$, then

$$F(x) = 0, \qquad\qquad\qquad \text{for } x < 1,$$

$$F(x) = \tfrac{1}{6}, \qquad\qquad\qquad \text{for } 1 \leq x < 2,$$

$$F(x) = \tfrac{1}{6} + \tfrac{1}{3} = \tfrac{1}{2}, \qquad \text{for } 2 \leq x < 3,$$

$$F(x) = 1, \qquad\qquad\qquad \text{for } 3 \leq x.$$

The distribution function is an important concept in both the discrete and continuous cases. There is a simple relationship between the probability function and the distribution function in either case. In the discrete case the relationship is as follows. Suppose the discrete random variable X has possible values $x_1, x_2, \ldots, x_i, \ldots$, arranged in order of increasing value. Then it is clear from the definition above that

$$p(x_1) = P(X = x_1) = F(x_1),$$

$$p(x_i) = P(X = x_i) = F(x_i) - F(x_{i-1}), \qquad i \geq 2.$$

Example 1 (continued). Reversing the example above with $x_1 = 1$, $x_2 = 2$, and $x_3 = 3$, we have

$$p(1) = P(X = 1) = F(1) = \tfrac{1}{6} = \tfrac{1}{6},$$

$$p(2) = P(X = 2) = F(2) - F(1) = \tfrac{1}{2} - \tfrac{1}{6} = \tfrac{1}{3},$$

$$p(3) = P(X = 3) = F(3) - F(2) = 1 - \tfrac{1}{2} = \tfrac{1}{2}.$$

PROPERTIES OF THE DISTRIBUTION FUNCTION. The distribution function is a nondecreasing function and, for a discrete random variable X, its graph is a series of horizontal lines with vertical jumps between them at the values $x_1, x_2, \ldots, x_i, \ldots$, which X can take. The size of the first jump is $F(x_1) = p(x_1)$ and the size of the ith jump for $i \geq 2$ is $F(x_i) - F(x_{i-1}) = p(x_i)$. Thus the size of the jump at x_i is the probability that X has the value x_i at which the jump occurs. The value of the distribution function at $-\infty$, $F(-\infty)$, is zero, and its value at $+\infty$, $F(+\infty)$, is one.

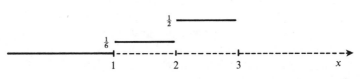

Figure 5.1.   Distribution function for Example 1.

Example 1 (continued).   The graph of the distribution function for the example given above is shown in Figure 5.1.

Example 2.   The probability function for the distribution of the number of heads obtained in three tosses of a balanced coin is

$$p(0) = \tfrac{1}{8}, \quad p(1) = \tfrac{3}{8}, \quad p(2) = \tfrac{3}{8}, \quad p(3) = \tfrac{1}{8}.$$

The corresponding distribution function is

$$F(x) = 0, \qquad x < 0,$$
$$F(x) = \tfrac{1}{8}, \qquad 0 \le x < 1,$$
$$F(x) = \tfrac{1}{2}, \qquad 1 \le x < 2,$$
$$F(x) = \tfrac{7}{8}, \qquad 2 \le x < 3,$$
$$F(x) = 1, \qquad 3 \le x.$$

The graph of this distribution function is shown in Figure 5.2.

Example 3.   Suppose the random variable X has a distribution with probability function given by

$$p(x) = 1/2^x, \qquad \text{for } x = 1, 2, \ldots.$$

Then we can write the corresponding distribution function as

$$F(x) = 0, \qquad\qquad \text{for } x < 1,$$
$$F(x) = 1 - 1/2^{[x]}, \qquad \text{for } x \ge 1,$$

where $[x]$ denotes the integer part of x, that is, the largest integer that is

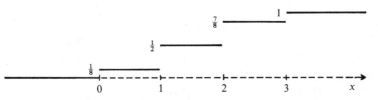

Figure 5.2.   Distribution function for the number of heads in three tosses of a balanced coin.

contained in the value given to x; for example, if we set $x = 3.4$, $[x] = 3$. The graph of this distribution function is especially interesting. The jumps of height $1/2^x$ at $x = 1, 2, \ldots$ get smaller and smaller, and $F(x)$ gets closer and closer to one, as x gets larger and larger. However, $F(x)$ reaches one only at infinity. (The reader should now draw the graph of this distribution function for himself.)

If we begin with the distribution function, we can recover the probability function as follows:

$$p(1) = F(1) = (1 - \tfrac{1}{2}) = \tfrac{1}{2},$$

$$p(2) = F(2) - F(1) = (1 - \tfrac{1}{4}) - (1 - \tfrac{1}{2}) = \tfrac{1}{4},$$

and, in general,

$$p(x) = F(x) - F(x - 1) = (1 - 1/2^x) - (1 - 1/2^{x-1}) = 1/2^x.$$

THE DISTRIBUTION FUNCTION FOR THE BINOMIAL DISTRIBUTION.   We have

$$F(x) = \sum_{t \leq x} \binom{n}{t} p^t (1 - p)^{n-t} \qquad \text{where } t \text{ takes integer values only.}$$

Note. The distribution function above is a function that gives us the probability of "x or fewer" successes. The cumulative binomial probability defined in Section 4.4, and tabulated for some n, p, and r in Table 4.1, gives us the probability of "r or more" successes. Note that, if we write $q = 1 - p$,

$$F(x) = \sum_{t \leq x} \binom{n}{t} p^t (1 - p)^{n-t} = \sum_{t \leq x} \binom{n}{n - t} (1 - p)^{n-t} p^t$$

$$= \sum_{t \leq x} \binom{n}{n - t} q^{n-t} (1 - q)^t$$

$$= \sum_{s > n - x} \binom{n}{s} q^s (1 - q)^{n-s}.$$

In words, the distribution function $F(x)$ for a binomial distribution with parameters n and p is equal to the cumulative binomial probability B_{n-x} for a binomial distribution with parameters n and $q = 1 - p$. We have had an example of this in Exercise 25(b) on page 101.

Example.   Ten coins are tossed. What is the probability of getting four or fewer heads? The answer is (where p is the probability of getting a head on a single toss)

$$P(X \le 4) = F(4) = \sum_{t=0}^{4} \binom{10}{t} p^t (1 - p)^{10-t}$$

$$= \binom{10}{0} p^0 (1 - p)^{10} + \binom{10}{1} p^1 (1 - p)^9 + \binom{10}{2} p^2 (1 - p)^8$$

$$+ \binom{10}{3} p^3 (1 - p)^7 + \binom{10}{4} p^4 (1 - p)^6.$$

If the coin is balanced so that $p = 1 - p = \frac{1}{2}$, then

$$F(4) = (\tfrac{1}{2})^{10}\{1 + 10 + 45 + 120 + 210\} = \tfrac{193}{512}.$$

Exercises

1. What are the distribution functions for the following random variables?:
   (a) A basketball player makes 80% of the free throws he attempts. Let X = the number of free throws made in 10 attempts.
   (b) A coin is biased 2 to 1 in favor of heads. Let X = the number of heads in eight tosses.
2. Draw a graph for each of the distribution functions in Exercise 1.
3. What are the distribution functions for the following random variables?:
   (a) X is geometric; $p(x) = (1 - p)^{x-1} p$, $x = 1, 2, \ldots$.

   (b) X is Pascal; $p(x) = \binom{x - 1}{r - 1} p^r (1 - p)^{x-r}$, $x = r, r + 1, r + 2, \ldots$.

   (c) X is negative binomial; $p(x) = \binom{r + x - 1}{x} p^r (1 - p)^x$,

   $x = 0, 1, 2, \ldots$.
4. (a) For the distribution of Exercise 3(a), find $F(3)$ when $p = \frac{1}{4}$.
   (b) For the distribution of Exercise 3(b), find $F(4)$ when $r = 3, p = \frac{1}{3}$.
   (c) For the distribution of Exercise 3(c), find $F(2)$ when $r = 4, p = \frac{1}{4}$.
5. The distribution function of a random variable is given by
   $F(x) = 0$,   for $x < 1$,
   $F(x) = x(2 + x)/63$,   at $x = 1, 2, \ldots, 7$,
   $F(x) = F([x])$,   for $1 \le x \le 7$, where $[x]$ denotes the integer part of x,
   $F(x) = 1$,   for $x > 7$.
   (a) Draw a graph of the distribution function $F(x)$.
   (b) Find the formula for $p(x)$.
   (c) Is $\sum p(x) = 1$?
   (d) Evaluate $p(x)$ for all values of x.
   (e) Do the numerical values you obtained in (d) add to one?
   (f) Draw a graph of the probability function $p(x)$.

6. Repeat (a) to (f) of Exercise 5 when

$F(x) = 0$,	for $x < 1$,
$F(x) = x(1 + x)/90$,	at $x = 1, 2, \ldots, 9$,
$F(x) = F([x])$,	for $1 \leq x \leq 9$, where $[x]$ denotes the integer part of x,
$F(x) = 1$,	for $x > 9$.

7. If X has a geometric distribution $p(x) = (1 - p)^{x-1}p$, $x = 1, 2, \ldots$, find $P(X > a + b | X > a)$. The result illustrates the property that the geometric distribution "has no memory."

5.2. A Continuous Random Variable: The Uniform Distribution

The notion of a distribution function provides a convenient approach to the study of continuous random variables. Although a complete treatment of continuous distributions requires a knowledge of calculus (which we assume the reader does not have), we may nevertheless obtain some idea of the differences between the discrete and continuous types of distribution without calculus. By limiting our discussion to the most simple example of a continuous random variable, one which has the so-called *uniform distribution*, we shall not get into any mathematical difficulties.

Suppose an 8-inch horizontal line is drawn across a sheet of paper 8 inches wide and then, in a random manner, another line is drawn on the paper, intersecting the horizontal line in (of course) one point. The distance of the point of intersection from the left edge of the paper is then a random variable X whose set of possible values is the set of real numbers between 0 and 8, $\{x : 0 < x < 8\}$. This random variable is not discrete, but continuous, because the set of possible values is noncountably infinite. (The actual boundary values 0 and 8 may, or may not, be included in the set of possible values without changing any of the following discussion.)

What is the probability that X will assume a particular value x? Intuitively, if we assume that each one of the infinite set of values of x has the same chance of being the point of intersection, it seems that the probability should be very close to zero. That it is, in fact, zero will be a consequence of the definition of probability for this random variable.

It seems reasonable to require the probability that the point of intersection will be on the left half of the sheet to be $\frac{1}{2}$, that it will be on the left quarter of the sheet to be $\frac{1}{4}$, that it will be on the left three-fourths of the sheet to be $\frac{3}{4}$, and so on. This suggests the following definition.

Definition.   If the sample space S of a continuous random variable X is the interval of real numbers from a to b, and if X is "uniformly distributed" over the interval, then the distribution function of X will be defined by

$$F(x) = \frac{x - a}{b - a}.$$

The numerator, $x - a$, is the distance from the left edge of the interval and the denominator is the size of the interval; then the fraction represents the probability that X will assume a value less than or equal to x. It is clear that F is an increasing function and that $F(a) = 0$ and $F(b) = 1$. The graph of $F(x)$ is shown in Figure 5.3.

We may obtain the expression for the probability function of X by using the obvious relationship between the distribution function and the probability function, namely that the probability that X will assume a value within a particular interval, say the interval from c to d, is given by $F(d) - F(c)$. Thus

$$P(c < X < d) = F(d) - F(c) = \frac{d - a}{b - a} - \frac{c - a}{b - a} = \frac{d - c}{b - a}.$$

It follows that the probability that X will assume the *particular value* x is given by the fraction whose numerator is $x - x$, and so is zero.

The probability that X will assume a value in any of several nonoverlapping intervals is given by the sum of fractions for the intervals, for each interval represents an event, and nonoverlapping intervals represent mutually exclusive events. Using the example about the 8-inch line at the beginning of this section, for instance, the probability that the point of intersection will be within 1 inch of either side of the sheet is given by

$$\frac{1 - 0}{8 - 0} + \frac{8 - 7}{8 - 0} = \frac{2}{8 - 0} = \frac{1}{4}.$$

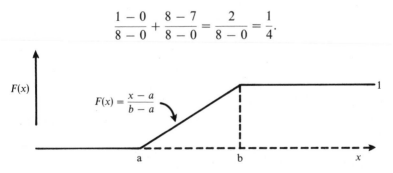

$$F(x) = \frac{x - a}{b - a}$$

Figure 5.3.   Distribution function for a uniform random variable.

It is seen that the fraction $2/(8 - 0)$ has as numerator the total length of intervals belonging to the event, and the denominator is the total domain of the function. If the intervals are *overlapping*, the required probability that S will assume a value in any of the intervals is found by taking as numerator the length of the union of the intervals belonging to the event. For instance, the probability that S in the previous example will be in any one of the intervals $(0, 2)$, $(1, 3)$, or $(2.5, 5)$ [where, for example, $(1, 3)$ means the interval between 1 inch and 3 inches from the left edge] is given by $\frac{5}{8}$, because the union of the three intervals is the interval $(0, 5)$.

Example 1. Consider the choice at random of a point on the circumference of a circle. The probability that the point will be on a particular arc, say between $c°$ and $d°$, is given by $(d - c)/360$. Thus the probability that the point will be between 135° and 180° measured from some fixed point is $(180 - 135)/360 = \frac{1}{8}$. The distribution function is given by $F(x) = x/360$, where x is measured in degrees from zero. The probability that the point will be not more than 120° from the 0° point is $\frac{120}{360} = \frac{1}{3}$.

Example 2. A slightly different example is the following. Suppose that buses leave a particular stop every 10 minutes between 7:00 and 8:00 A.M. beginning at 7:00. If a student's time of arrival at the bus stop is a uniform random variable with possible values between 7:00 and 8:00, what is the probability that he will have to wait more than 5 minutes for a bus?

We can divide the sample space into subintervals as follows: 7:00–7:05, 7:05–7:10, ..., 7:55–8:00. The student will have to wait more than 5 minutes if he arrives at the stop between 7:00 and 7:05, 7:10 and 7:15, ..., 7:50 and 7:55, so that the probability of waiting more than 5 minutes is the ratio of the sum of these intervals to the whole interval of 60 minutes: $\frac{30}{60} = \frac{1}{2}$.

Exercises

1. Find the probability that a driver will have a green light at an intersection if his time of arrival is by chance and the light alternates from 30 seconds of green to 20 seconds of red.
2. If X is uniformly distributed over the interval from -1 to $+2$, find
   (a) $P(X > 0)$.   (c) $P(0 < X < 1)$.
   (b) $P(X < \frac{1}{2})$.   (d) $P(X < 0 | X < 1)$.
3. A point is chosen at random on a line segment of length 5 inches. Find the probability that it is within 1 inch of the center.
4. A real number is chosen at random from the interval $(5, 10)$. What is the probability that
   (a) it is between 7 and 8.5?
   (b) it is less than 9.5?
   (c) it is exactly 7.5?

5. The time of arrival of a secretary at her office is a uniform random variable taking any value between 8:10 and 8:25. Find the probability that on a particular day she will arrive
    (a) before 8:20.
    (b) between 8:20 and 8:25.
    (c) at exactly 8:20.
6. The 6:00 A.M. temperature at a certain place in the West Indies is a uniform random variable with values between 67 and 74°F. Find the probability that on a particular day the 6:00 A.M. temperature will be
    (a) under 70°F.   (b) over 72°F.   (c) between 69 and 71°F.
7. The time interval between customers in a furniture store is a uniform random variable with values between 2 minutes and $1\frac{1}{2}$ hours. The salesman wants to go next door for coffee. Find the probability that there will be no customer during the 15-minute interval that he will be gone.
8. Apply the principle of a uniform distribution to the following problem. On a piece of paper 8 by $10\frac{1}{2}$ inches, four squares 1 by 1 inch each are drawn without overlapping. Find the probability that a point chosen at random on the paper will fall in a square.

5.3. Expected Value, Mean, and Variance

INTRODUCTION. The notion of *expected value* (or *mathematical expectation*) can be illustrated by examples from games of chance.

Example 1. Suppose two players, A and B, play a series of games as follows: Three fair coins are tossed simultaneously. B is the banker; he receives $5 from A for each game and he pays A $3 for each head that appears. What average amount per game will A win if he plays this game many times? That is, what is the expected value in money of a game to A?

To answer this question we must examine A's chances of winning the various amounts he can win and the amount he pays to play. There are four possible outcomes for A: 0 heads, 1 head, 2 heads, and 3 heads. These are not equally probable; in fact, since the number of heads is a binomial variable, we can show that $p(0) = \frac{1}{8}$, $p(1) = p(2) = \frac{3}{8}$, and $p(3) = \frac{1}{8}$. Since A

pays \$5 to participate in each game, he can thus win

$$\$0 - \$5 = -\$5 \text{ with probability } \tfrac{1}{8},$$

$$\$3 - \$5 = -\$2 \text{ with probability } \tfrac{3}{8},$$

$$\$6 - \$5 = \$1 \text{ with probability } \tfrac{3}{8},$$

$$\$9 - \$5 = \$4 \text{ with probability } \tfrac{1}{8}.$$

We define his *expected* win per game in dollars as

$$(-5)(\tfrac{1}{8}) + (-2)(\tfrac{3}{8}) + (1)(\tfrac{3}{8}) + (4)(\tfrac{1}{8}) = -0.5.$$

(From a frequency point of view we can argue that, in a long series of games, A should lose \$5 in $\tfrac{1}{8}$ of the games, lose \$2 in $\tfrac{3}{8}$ of the games, win \$1 in $\tfrac{3}{8}$ of the games, and win \$4 in $\tfrac{1}{8}$ of the games, giving him an average loss per game of \$0.50, through the same arithmetic.) In other words, A's average win is $-\$0.50$; that is, he actually loses \$0.50 per game on the average. This is the *expected value* in money of the game as far as A is concerned. (The expected value in money for B is, in this case, a gain of \$0.50 per game.) Clearly, A would be well advised not to participate in this game!

Example 2.   Suppose we remove the financial aspects from the game in Example 1 and ask, instead, how many heads we can expect to get per game. We want now the expected value of *the number of heads* (and not of the money won). By similar arguments to those above, we see we shall obtain

0 heads with probability $\tfrac{1}{8}$,

1 head with probability $\tfrac{3}{8}$,

2 heads with probability $\tfrac{3}{8}$,

3 heads with probability $\tfrac{1}{8}$,

giving the expected value of the number of heads per game (or the expected number of heads) as

$$0(\tfrac{1}{8}) + 1(\tfrac{3}{8}) + 2(\tfrac{3}{8}) + 3(\tfrac{1}{8}) = \tfrac{12}{8} = 1.5 \text{ heads}.$$

(Again, from a frequency point of view, we should expect, in a long series of games, 0 heads in $\tfrac{1}{8}$ of the games, 1 head in $\tfrac{3}{8}$ of the games, 2 heads in $\tfrac{3}{8}$ of the games, and 3 heads in $\tfrac{1}{8}$ of the games. The average number of heads per game is thus 1.5, by the same arithmetic as above.)

In both of the foregoing examples we found expected values of random variables which had probability distributions. The answers told us something about each distribution—what we could expect "on the average" from observations of the corresponding random phenomenon. Note that, in both cases, the expected value was *not* a number that could actually be obtained from a single trial; this is not unusual, in general.

We now define the term expected value more generally and consider some of the quantities which can be obtained from this definition and which we shall later find of value in describing distributions.

EXPECTED VALUE OF A FUNCTION OF A DISCRETE RANDOM VARIABLE. Let X be a discrete random variable with probability function $p(x)$. Let $g(X)$ be any function of X. Then the expected value of $g(X)$ is defined as follows:

$$Eg(X) = E(g(X)) = \sum_x g(x)p(x),$$

the summation being taken over all possible values of x. The capital E denotes "expectation of" or "expected value of" and can be written with or without parentheses, as shown.

EXPECTED VALUE OF A CONSTANT. Set $g(x) = d$, a constant. Then

$$Ed = \sum_x dp(x)$$

$$= d \sum_x p(x)$$

$$= d,$$

since $\sum_x p(x) = 1$.

MEAN OF A DISCRETE DISTRIBUTION. Set $g(X) = X$. Then the *mean* of a distribution, its average value, is defined as

$$\mu = EX = E(X) = \sum_x x p(x).$$

It is usually denoted by the Greek letter μ (mu) as indicated. Again, the notations EX and $E(X)$ are interchangeable and equivalent. Note that the mean is a weighted sum of the possible values of X, each value being weighted by the probability of its occurrence. If we consider the probability as a relative frequency, that is, as the ratio of the frequency of occurrence of the value x to the total number of observations of the random phenomenon, then EX can be interpreted as a weighted average of the values of X, each value being weighted by its frequency of occurrence in n observations. Thus, if f_i is the frequency of occurrence of the value x_i and if the probability of x_i is taken to be f_i/n, then

$$EX = \sum_i x_i \frac{f_i}{n} = \frac{1}{n} \sum_i x_i f_i.$$

Example 1.   Consider the distribution in which X can take the three values $x = -1, 0$, and 1 with probabilities

$$p(-1) = \tfrac{1}{6}, \quad p(0) = \tfrac{1}{3}, \quad p(1) = \tfrac{1}{2}.$$

The mean is

$$\mu = \sum_x x\, p(x) = -1(\tfrac{1}{6}) + 0(\tfrac{1}{3}) + 1(\tfrac{1}{2}) = \tfrac{1}{3}.$$

Example 2   (Mean of a Binomial Variable X). For a binomial variable

$$\mu = EX = \sum_{x=0}^{n} x\binom{n}{x} p^x (1 - p)^{n-x}$$

$$= \sum_{x=0}^{n} \frac{x\, n!}{x!(n-x)!} p^x (1 - p)^{n-x}$$

$$= np \sum_{x=1}^{n} \frac{(n-1)!}{(x-1)!(n-x)!} p^{x-1} (1 - p)^{n-x}.$$

We now set $s = x - 1$. As x goes from 1 to n, s goes from 0 to $n - 1$. Thus we have, continuing,

$$\mu = np \sum_{s=0}^{n-1} \binom{n-1}{s} p^s (1 - p)^{n-1-s}$$

$$= np.$$

This last step follows because the summation is simply the expansion of $\{p + (1 - p)\}^{n-1} = 1$. Thus the mean of a binomial variable is $\mu = np$.

MEAN AS CENTER OF GRAVITY.   The mean can be thought of, also, as the center of gravity of the distribution; for, if we regard each "spike" on the graph of a discrete distribution as a weight equal to the probability assigned to the corresponding value of X, then the mean of the distribution is the center of gravity of the weights, by definition. It is the value of X at which the system of weights would balance if they were supported by a weightless plank. Note that this point of balance is not necessarily a value that X actually takes.

Example.   Suppose X takes the value -1 with probability $\tfrac{1}{6}$, the value 0 with probability $\tfrac{1}{3}$, and the value 1 with probability $\tfrac{1}{2}$. The mean is $-1(\tfrac{1}{6}) + 0(\tfrac{1}{3}) + 1(\tfrac{1}{2}) = \tfrac{1}{3}$. Thus the weights $\tfrac{1}{6}$ at -1, $\tfrac{1}{3}$ at 0, and $\tfrac{1}{2}$ at 1 would balance at the point $\tfrac{1}{3}$, as shown in Figure 5.4.

A USEFUL RULE.   It is easy to see that if $g(X)$ and $h(X)$ are any functions of a random variable X with probability distribution function $p(x)$, then

$$E\{g(X) + h(X)\} = Eg(X) + Eh(X).$$

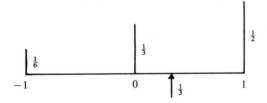

Figure 5.4. Mean as center of gravity or "point of balance."

This is because the left-hand side can be written

$$\sum_x \{g(x) + h(x)\}p(x) = \sum_x \{g(x)p(x) + h(x)p(x)\}$$
$$= \sum_x g(x)p(x) + \sum_x h(x)p(x)$$
$$= \text{right-hand side.}$$

Example. Let $g(X) = X$ and $h(X) = d$, a positive or negative real constant. Then

$$E(X + d) = EX + Ed = EX + d.$$

Thus, if a random variable (and hence the distribution of a random variable) is shifted by an amount d, so is its mean value.

To illustrate this, consider the variable X that takes the values $x = (-1, 0, 1)$ with probabilities $p(x) = (\frac{1}{6}, \frac{1}{3}, \frac{1}{2})$, respectively. We know that $EX = \frac{1}{3}$ from above. Choose $d = 3$ and let $Z = X + 3$; then Z takes the values $z = [(-1 + 3), (0 + 3), (1 + 3)]$, that is, the values $z = (2, 3, 4)$, with probabilities $p(z) = (\frac{1}{6}, \frac{1}{3}, \frac{1}{2})$, as in Figure 5.5. We have

$$EZ = 2(\tfrac{1}{6}) + 3(\tfrac{1}{3}) + 4(\tfrac{1}{2}) = 3\tfrac{1}{3} = \tfrac{1}{3} + 3 = EX + d.$$

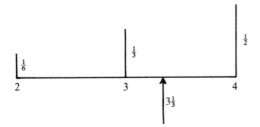

Figure 5.5. Mean of a distribution shifted by three units. (Compare with Figure 5.4.)

MOMENTS OF A DISCRETE DISTRIBUTION ABOUT A GENERAL POINT a.   Setting $g(X) = (X - a)^k$, we define

$$E(X - a)^k = \sum_x (x - a)^k p(x)$$

as the kth moment about the point a, for any discrete distribution.

Special case (i).   Set $a = 0$. Then we have the kth moment of the distribution about zero given by

$$\mu'_k = EX^k = \sum_x x^k p(x).$$

Special case (ii).   Set $a = 0$, $k = 1$. We obtain the mean

$$\mu = \mu'_1 = EX = \sum_x x p(x).$$

Special case (iii)   (moments about the mean). Set $a = \mu$, the mean of the distribution. Then we have the kth moment of the distribution about the mean given by

$$\mu_k = E(X - \mu)^k = \sum_x (x - \mu)^k p(x).$$

Special case (iv)   (variance). Set $a = \mu$, $k = 2$. We obtain the second ($k = 2$) moment of the distribution about the mean, called the *variance* and usually denoted by the Greek σ^2 (sigma squared), or by $V(X)$, namely

$$V(X) = \sigma^2 = E(X - \mu)^2 = \sum_x (x - \mu)^2 p(x).$$

The variance tells us something about the spread, or dispersion, of the discrete distribution about its mean; it tells us something additional about the allocation of probabilities to the values of the random variable. If the values close to the mean have the highest probabilities (we say "are most probable"), then the variance will be relatively low because the largest values of $p(x)$ will be multiplied by small values of $(x - \mu)^2$. If the values at a greater distance from the mean have higher probabilities than those close to the mean, then the variance will be large because now we will have large values of $p(x)$ multiplied by large values of $(x - \mu)^2$ (see Figure 5.6).

The information given by the variance about a distribution is different from that given by the mean (which is a measure of location) and can help in distinguishing between two distributions having the same mean. The distributions shown in Figure 5.6 have the same mean, zero, and, in fact,

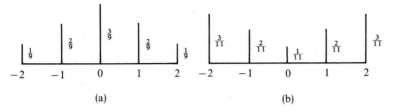

Figure 5.6. Distributions with the same mean and the same x values but different variances.

also the same x values. If we now work out the variances, however, we obtain, respectively:

(a) $\sigma^2 = (-2 - 0)^2(\frac{1}{9}) + (-1 - 0)^2(\frac{2}{9}) + (0 - 0)^2(\frac{1}{3}) + (1 - 0)^2(\frac{2}{9})$
$+ (2 - 0)^2(\frac{1}{9}) = \frac{4}{3}$.

(b) $\sigma^2 = (-2 - 0)^2(\frac{3}{11}) + (-1 - 0)^2(\frac{2}{11}) + (0 - 0)^2(\frac{1}{11})$
$+ (1 - 0)^2(\frac{2}{11}) + (2 - 0)^2(\frac{3}{11}) = \frac{28}{11}$.

These values reflect the facts that distribution (a) has its higher probabilities close to the mean, and distribution (b) has its higher probabilities farther from the mean.

Since the variance is a measure of spread about the mean of a distribution, it ought *not* to be affected by shifting the whole distribution to the right or the left. That this is indeed true can be shown as follows. Suppose the random variable X is shifted by an amount d to $(X + d)$, where d could be positive or negative. Then by definition of variance

$$V(X + d) = E\{X + d - E(X + d)\}^2$$
$$= E\{X + d - EX - d\}^2 \qquad \text{(p. 115)}$$
$$= E\{X - EX\}^2$$
$$= V(X);$$

that is, the shift has left the variance unchanged, as expected.

THE STANDARD DEVIATION. Sometimes it is more convenient to work with the positive square root of the variance, that is, with σ rather than with σ^2, because the units of σ are the same as those of X, whereas σ^2 has squared units. This measure, σ, is called the *standard deviation* of the distribution, or simply the standard deviation of the variable X.

Example. Find the variance and standard deviation of the distribution in which X can take the three values $x = -1, 0$, and 1 with probabilities

$$p(-1) = \tfrac{1}{6}, \quad p(0) = \tfrac{1}{3}, \quad p(1) = \tfrac{1}{2}.$$

We recall from Example 1 that the mean is $\frac{1}{3}$. Hence

$$V(X) = \sum (x - \mu)^2 p(x)$$
$$= (-1 - \tfrac{1}{3})^2(\tfrac{1}{6}) + (0 - \tfrac{1}{3})^2(\tfrac{1}{3}) + (1 - \tfrac{1}{3})^2(\tfrac{1}{2})$$
$$= \tfrac{5}{9}.$$

The standard deviation is the positive square root: $\sigma = \frac{1}{3}\sqrt{5}$.

THE VARIANCE: AN ALTERNATIVE FORMULA. We can write

$$V(X) = \sigma^2 = \sum_x (x - \mu)^2 p(x)$$
$$= \sum_x (x^2 - 2\mu x + \mu^2)p(x)$$
$$= \sum_x x^2 p(x) - 2\mu \sum_x xp(x) + \mu^2 \sum p(x).$$

Since $\sum xp(x) = \mu$ and $\sum p(x) = 1$, we obtain

$$V(X) = \sigma^2 = \sum_x x^2 p(x) - 2\mu^2 + \mu^2$$
$$= \sum_x x^2 p(x) - \mu^2.$$

That is,

$$V(X) = EX^2 - (EX)^2,$$

an alternative formula which is sometimes more convenient for computation.

Example.   Reworking the most recent example we have

$$\mu^2 = (EX)^2 = \tfrac{1}{9},$$
$$EX^2 = (-1)^2 \cdot \tfrac{1}{6} + (0)^2 \cdot \tfrac{1}{3} + (1)^2 \cdot \tfrac{1}{2} = \tfrac{2}{3} = \tfrac{6}{9},$$
$$V(X) = EX^2 - (EX)^2 = \tfrac{5}{9},$$

as above.

A USEFUL TRICK.   The fact that $E[g(x) + h(x)] = Eg(x) + Eh(x)$ can often be turned to advantage. For example, it means we can write

$$EX(X - 1) = E(X^2 - X) = EX^2 - EX.$$

We shall use this in the example below in the following way:

$$V(X) = EX^2 - (EX)^2$$
$$= EX^2 - EX + EX - (EX)^2$$
$$= EX(X - 1) + EX - (EX)^2.$$

Example. (Variance of a Binomial Variable X.) First, we can show that
$$EX(X - 1) = n(n - 1)p^2.$$
We leave the full detail of this calculation as an exercise for the reader. Note the trick here. We work with $EX(X - 1)$ and not with EX because the former allows us to cancel out two factors $x(x - 1)$ in the denominator of the $x!$ which appears in $p(x)$; that is,

$$EX(X - 1) = n(n - 1)p^2 \sum_{x=2}^{n} \binom{n - 2}{x - 2} p^{x-2}(1 - p)^{n-x},$$

and, similarly to the mean calculation, the summation is $\{p + (1 - p)\}^{n-2} = 1$. Now we recall that $EX = np$ and use the formula given immediately above this example to write

$$\begin{aligned} V(X) &= n(n - 1)p^2 + np - n^2p^2 \\ &= n^2p^2 - np^2 + np - n^2p^2 \\ &= np(1 - p). \end{aligned}$$

ANOTHER USEFUL RULE.   It is also easy to see that if $g(X)$ is any function of a random variable X with probability function $p(x)$, and if c is any constant, then
$$E\{cg(X)\} = cEg(X).$$

Example 1. $E(cX) = cEX = c\mu.$

Example 2. $E(cX + d) = E(cX) + d$ (pp. 114 and 113)
$$= cE(X) + d.$$

Example 3. $\begin{aligned} V(cX) &= E(c^2 X^2) - \{E(cX)\}^2 \\ &= c^2 EX^2 - \{cE(X)\}^2 \\ &= c^2\{EX^2 - (EX)^2\} \\ &= c^2 V(X). \end{aligned}$

(For an application see page 174, with $c = 1/n$.)

Example 4. $\begin{aligned} V(cX + d) &= V(cX) \\ &= c^2 V(X). \end{aligned}$ (p. 117)

RECAPITULATION ON MEAN AND VARIANCE.   If X is a random variable with probability function $p(x)$, then

$$\text{mean} = EX = \mu = \sum_{x} xp(x),$$

$$\text{variance} = V(X) = EX^2 - (EX)^2 = \sigma^2 = \sum_{x} x^2 p(x) - \left\{\sum_{x} xp(x)\right\}^2.$$

For the binomial distribution $p(x) = \binom{n}{x} p^x (1 - p)^{n-x}$, we find $EX = np$, $V(X) = np(1 - p)$.

Exercises

1. A random variable takes the values 2, 3, 4, 7, and 8 with probabilities 0.12, 0.24, 0.20, 0.28, and 0.16, respectively. Find the mean and variance of the distribution.

2. Find the mean and standard deviation of the distribution in which X takes each of the values 1, 2, 3, 4, and 5 with probability 0.2.

3. For the distributions in Exercises 1 and 2 find
   (a) $E(3X)$.   (b) $E(2X + 1)$.   (c) $V(3X)$.   (d) $V(2X + 1)$.

4. If X has the probability function $p(x) = x/21$, $x = 1, 2, \ldots, 6$, find the mean and variance.

5. If X has the probability function $p(x) = x/45$, $x = 1, 2, \ldots, 9$, find the mean and standard deviation.

6. For the distributions in Exercises 4 and 5 find
   (a) $E(\frac{1}{2}X)$.   (b) $E(3X - 2)$.   (c) $V(\frac{1}{2}X)$.   (d) $V(3X - 2)$.

7. Find EX and $V(X)$ when X is the number of spots on the upper face of a fair die that is tossed randomly.

8. Find EX and $V(X)$ when X is the sum of the numbers obtained when two dice are tossed.

9. Find the mean and standard deviation of the distribution $p(0) = 1 - p$, $p(1) = p$.

10. Find EX and $V(X)$ when X is the number of heads in three tosses of a balanced coin.

11. Find the mean and standard deviation for the distribution of the number of heads in six tosses of a balanced coin.

12. Find EX and $V(X)$ for the variable X = the number of free throws made in 10 attempts by a basketball player who makes 80% of his attempts.

13. Find the mean and standard deviation for the distribution of the number of successes in 111 trials if the probability of success on a single trial is
    (a) $\frac{1}{3}$.   (b) $\frac{1}{37}$.

14. If the probability of success is 0.25 in a binomial distribution, find the mean and standard deviation of the number of successes when n is
    (a) 4.   (b) 8.   (c) 16.   (d) 32.   (e) 64.   (f) 128.
    What do you notice?

15. A coin is biased 2 to 1 in favor of heads. Let X be the number of heads in 8 tosses. Find EX and $V(X)$.

16. Find EX and $V(X)$ for the geometric distribution
    $$p(x) = (1 - p)^{x-1} p \qquad x = 1, 2, \ldots,$$
    where X is the number of trials needed to the first success. [*Answers*: $1/p$ and $(1 - p)/p^2$.]

17. Find EX and $V(X)$ for the Pascal distribution

$$p(x) = \binom{x-1}{r-1} p^r(1-p)^{x-r} \qquad x = r+1, r+2, \ldots,$$

where X is the number of trials required to obtain r successes. [*Answers*: r/p and $r(1-p)/p^2$.]

18. Find EX and $V(X)$ for the negative binomial distribution

$$p(x) = \binom{r+x-1}{x} p^r(1-p)^x \qquad x = 0, 1, 2, \ldots,$$

where X is the number of failures before the rth success. [*Answers*: $r(1-p)/p$ and $r(1-p)/p^2$.]

19. Show the details of the work in finding that the variance of the binomial distribution $p(x) = \binom{n}{x} p^x(1-p)^{n-x}$ is $np(1-p)$.

20. Find μ_1', μ_2', μ_3', and μ_4' for the distributions in Exercises 7, 9, 10, and 15.

21. Show that, for any discrete distribution, $\mu_2 = \mu_2' - \mu_1'^2$. Have we had this before?

22. Show that $\mu_3 = \mu_3' - 3\mu_1'\mu_2' + 2\mu_1'^3$.

23. Show that $\mu_3' = \mu_3 + 3\mu_1'\mu_2 + \mu_1'^3$.

24. Show that $\mu_4 = \mu_4' - 4\mu_1'\mu_3' + 6\mu_1'^2\mu_2' - 3\mu_1'^4$.

25. Show that $\mu_4' = \mu_4 + 4\mu_1'\mu_3 + 6\mu_1'^2\mu_2 + \mu_1'^4$.

26. For the binomial distribution find $EX(X-1)(X-2)$. Hence show that $\mu_3 = npq(q-p)$, where $q = 1 - p$. For what values of p is $\mu_3 = 0$? Is this obvious by definition of μ_3?

27. For the binomial distribution find $EX(X-1)(X-2)(X-3)$. Hence show that $\mu_4 = 3n^2p^2q^2 + npq(1-6pq)$, where $q = 1 - p$.

28. Find μ_1', μ_2', μ_3', and μ_4' for the geometric, Pascal, and negative binomial distributions in Exercises 16, 17, and 18. Hence find μ_1, μ_2, μ_3, and μ_4 via the results in Exercises 21, 22, and 24.

5.4. Chebyshev's Inequality

Chebyshev (whose name may be found spelled in various ways in various places) was a Russian mathematician (1821–1894). The Chebyshev inequality casts light on how the variance measures spread and can be stated in words as follows:

> For any random variable X with mean μ and variance σ^2, both finite, and for any positive number h, the probability of getting a value that differs from the mean in one direction or the other by h standard deviations or more is less than or equal to $1/h^2$.

In symbols we can write the inequality as

$$P(|X - \mu| \geq h\sigma) \leq 1/h^2.$$

Proof. The proof is not difficult. Let k be any positive number. Then we can write

$$\sigma^2 = \sum_{-\infty}^{\infty} (x - \mu)^2 p(x) = \sum_{-\infty}^{\mu-k} (x - \mu)^2 p(x)$$

$$+ \sum_{\substack{x > \mu-k \\ x < \mu+k}} (x - \mu)^2 p(x) + \sum_{\mu+k}^{\infty} (x - \mu)^2 p(x)$$

$$\geq \sum_{-\infty}^{\mu-k} (x - \mu)^2 p(x) + \sum_{\mu+k}^{\infty} (x - \mu)^2 p(x),$$

for the sum $\sum_{\substack{x > \mu-k \\ x < \mu+k}} (x - \mu)^2 p(x)$ is obviously nonnegative. Now, if $x \geq \mu + k$ or $x \leq \mu - k$, then $(x - \mu)^2 \geq k^2$. Therefore,

$$\sigma^2 \geq \sum_{-\infty}^{\mu-k} k^2 p(x) + \sum_{\mu+k}^{\infty} k^2 p(x) = k^2 \left[\sum_{-\infty}^{\mu-k} p(x) + \sum_{\mu+k}^{\infty} p(x) \right]$$

or,

$$\sigma^2 \geq k^2 \{ P(X \leq \mu - k) + P(X \geq \mu + k) \} = k^2 P((X - \mu)^2 \geq k^2)$$
$$= k^2 P(|X - \mu| \geq k).$$

It follows that

$$P(|X - \mu| \geq k) \leq \left(\frac{\sigma}{k} \right)^2,$$

so that if we write $h = k/\sigma$ we obtain the desired result,

$$P(|X - \mu| \geq h\sigma) \leq \frac{1}{h^2}.$$

(The proof above is appropriate for discrete distributions. For continuous distributions, the proof is of similar format, but there are notational changes that lie beyond the scope of this book.)

Note 1. Chebyshev's inequality usually is not a particularly tight one, as we shall see from an example in a moment, but this is not unreasonable, because absolutely nothing has been assumed about the distribution of X except that it has a mean and a variance! If we know more about the distribution—for example, that it is binomial—then we can always obtain more precise results. Even if more information is available, Chebyshev's inequality is still true, of course, but a tighter inequality can usually be obtained by employing the distributional information instead. The fact remains that Chebyshev's inequality tells us a great deal considering that the assumptions made to get it are minimal.

Example. Let $h = 2$. Then Chebyshev's inequality tells us that

$$P(|X - \mu| \geq 2\sigma) \leq \tfrac{1}{4}.$$

That is, the probability that an observation will fall two or more standard deviations away on either side of the mean is less than 0.25 for *any* distribution that has finite mean and variance.

Suppose we now assume, in addition, that X has a binomial distribution with parameters $p = \tfrac{1}{3}$ and $n = 9$. Then we can work out the exact probability as follows: $\mu = np = 3$, $\sigma^2 = np(1 - p) = 2$, so that $\sigma = 2^{1/2} = 1.414$.

$$P(|X - \mu| \geq 2\sigma) = P(X \leq \mu - 2\sigma) + P(X \geq \mu + 2\sigma)$$
$$= P(X \leq 0.172) + P(X \geq 5.828).$$

Since X can take only the discrete values $0, 1, 2, \ldots, 8, 9$, this is equal to

$$p(0) + p(6) + p(7) + p(8) + p(9)$$

$$= \left\{ \binom{9}{0}2^9 + \binom{9}{6}2^3 + \binom{9}{7}2^2 + \binom{9}{8}2 + \binom{9}{9} \right\} \left(\frac{1}{3}\right)^9$$

$$= \{528 + 504 + 144 + 18 + 1\}/3^9$$

$$= \frac{1195}{19683}$$

$$= 0.061.$$

We see, then, that the Chebyshev bound of 0.25 was, in fact, rather a loose one when applied to this *particular* distribution. However, as we remarked above, the Chebyshev result is true no matter what the distribution of X may be. To achieve such a general result we must expect to lose something in the tightness of the bound achieved on particular distributions.

Note 2. If we wish to make probability statements of the type above about a large class of random variables, then Chebyshev's inequality may be the best we can do.

Example. Suppose we consider the entire class of random variables X with mean μ and variance σ^2. One member of this class is the "three-point" variable X that can take the values $-h$, 0, and h ($h > 0$), with probabilities

$$p(-h) = p(h) = 1/2h^2 \qquad p(0) = 1 - 1/h^2.$$

Clearly, $EX = 0$ and $V(X) = 1$, so that

$$P(|X - \mu| \geq h\sigma) = P(|X| \geq h) = P(X = -h) + P(X = h) = 1/h^2.$$

Now Chebyshev's inequality gives

$$P(|X - \mu| \geq h\sigma) \leq 1/h^2,$$

but this limit is actually attained by one member of the class, namely the random variable given above. It follows that in this case Chebyshev's statement is the best possible about the probability concerned, for the whole class of random variables considered.

CHEBYSHEV'S INEQUALITY: COMPLEMENTARY STATEMENT.

We note that the inequality can be also stated in the complementary form

$$P(|X - \mu| < h\sigma) \geq 1 - \frac{1}{h^2}.$$

That is, the probability of getting a value *within h* standard deviations of the mean is at least $1 - 1/h^2$.

Exercises

1. State Chebyshev's inequality when $h = \sqrt{3}$. Now work out the exact probability on the left-hand side when the distribution of X is binomial with $n = 8$, $p = \frac{1}{4}$. Compare the two results.
2. State Chebyshev's inequality when $h = 3.5$. Now work out the exact probability on the left-hand side when the distribution of X is binomial with $n = 16$, $p = \frac{1}{2}$. Compare the two results.
3. State Chebyshev's inequality when $h = \sqrt{2}$. Now work out the exact probability on the left-hand side when the distribution of X is binomial with $n = 6$, $p = \frac{1}{2}$. Compare the two results.
4. Does Chebyshev's inequality apply when $h < 1$? Say yes or no, and discuss the usefulness of the inequality in this case.

CHAPTER 6

The Hypergeometric and Poisson Distributions

6.1. Hypergeometric Distribution

Suppose we have an urn that contains N balls consisting of W white balls and $N - W$ red balls. The probability of drawing a white ball (a "success") is clearly $p = W/N$, and the probability of drawing a red ball (a "failure") is $1 - p = (N - W)/N$, for a single trial. If we make n trials *with replacement*, that is, replacing the ball drawn on one trial and remixing the balls in the urn before the next trial, the number of white balls drawn, X, is a binomial variable with parameters n and p, so that

$$p(x) = \binom{n}{x} \left(\frac{W}{N} \right)^x \left(\frac{N - W}{N} \right)^{n-x} \qquad x = 0, 1, \ldots, n.$$

Suppose, however, that balls drawn are *not* replaced. Then the binomial law does not apply because the probability of a success is not constant from trial to trial. The appropriate probability law for X, the number of white balls drawn in $n \leq N$ trials, is then a *hypergeometric distribution* given by

$$p(x) = \frac{\binom{W}{x}\binom{N-W}{n-x}}{\binom{N}{n}} \qquad x = \begin{cases} 0, 1, \ldots, W & \text{for } n \geq W, \\ 0, 1, \ldots, n & \text{for } n < W. \end{cases}$$

We can argue as follows. There are $\binom{N}{n}$ equally likely ways of selecting n balls from the urn. There are $\binom{W}{x}$ equally likely ways of drawing x white balls from the W available and $\binom{N-W}{n-x}$ equally likely ways of drawing $n-x$ red balls from the $N-W$ available. There are, thus,

$$\binom{W}{x}\binom{N-W}{n-x}$$

ways of drawing x white and $n-x$ red balls, and the probability of drawing exactly x white balls in n draws is this product divided by the total number of possible ways of drawing n balls in the urn. Hence the result above.[1]

[1] A more detailed argument is as follows. The probability of a sequence of x white balls followed by $n-x$ red balls is, by the use of conditional probabilities,

$$\left(\frac{W}{N}\right)\left(\frac{W-1}{N-1}\right)\cdots\left(\frac{W-x+1}{N-x+1}\right)\left(\frac{N-W}{N-x}\right)\left(\frac{N-W-1}{N-x-1}\right)\cdots\left(\frac{N-W-n+x+1}{N-n+1}\right)$$

$$= \frac{W!(N-x)!}{(W-x)!N!}\frac{(N-W)!}{(N-W-n+x)!}\frac{(N-n)!}{(N-x)!}.$$

Multiplying both numerator and denominator by $x!(n-x)!$, we obtain

$$\frac{\binom{W}{x}\binom{N-W}{n-x}}{\binom{N}{x}\binom{N-x}{n-x}}.$$

Now there are $\dfrac{n!}{x!(n-x)!} = \dbinom{n}{x}$ different sequences of n balls of which x are white and $n-x$ are red, and each has the same probability. Hence the probability of the result "exactly x white balls and $n-x$ red balls" is

$$p(x) = \frac{\binom{n}{x}\binom{W}{x}\binom{N-W}{n-x}}{\binom{N}{x}\binom{N-x}{n-x}} = \frac{\binom{W}{x}\binom{N-W}{n-x}}{\binom{N}{n}},$$

after reduction.

The hypergeometric distribution applies in a great many situations. For example, if cards are drawn *without replacement* from a bridge deck, the binomial distribution no longer applies because the probability of success— of getting a spade, say—is not constant for all trials but changes with each trial and depends upon the cards drawn previously. In this case, the hypergeometric distribution is used with W representing the total number of spades in the deck, 13. In general, we may say that W represents the total number of objects with the characteristic labeled "success" among the N objects from which we are to draw a subset of size n.

Example. Suppose we draw a "hand" of 13 cards without replacement from a bridge deck. What is the probability $p(7)$ that the hand will contain 7 spades? First let us determine the denominator. We are drawing $n = 13$ cards from a set of size $N = 52$; we can do this in $\binom{52}{13}$ ways. The numerator is the product of the $\binom{13}{7}$ ways of drawing 7 spades from the 13 spades in the deck and the $\binom{39}{6}$ ways of drawing the other 6 cards from the 39 nonspades in the deck. Thus

$$p(7) = \frac{\binom{13}{7}\binom{39}{6}}{\binom{52}{13}} = 0.0088,$$

that is, the chance that seven spades will occur is less than 1%.

CALCULATING HYPERGEOMETRIC PROBABILITIES.

Owing to the size of the coefficients $\binom{m}{t}$ involved in many problems, the calculations needed to get hypergeometric probabilities cannot usually be done with the apparatus available to us so far in this book. [In Table 1.2, for example, we give values of $\binom{m}{t}$ only as far as $m = 20$.] We can, however, use Tables 6.1 and 6.2 in such situations. Table 6.1 gives four-figure logarithms to the base 10 for $\log m!$ when $1 \leq m \leq 100$. Table 6.2 is an abbreviated table in which we can look up the antilogarithm of a number correct to two figures (or to more if we linearly interpolate the other figures in the table). We illustrate the use of these tables to get $p(7)$ in the example above.

Table 6.1. Common Logarithms of Factorials[a]

m	$\log m!$	m	$\log m!$	m	$\log m!$	m	$\log m!$
1	0.0000	26	26.6056	51	66.1906	76	111.2754
2	0.3010	27	28.0370	52	67.9067	77	113.1619
3	0.7782	28	29.4841	53	69.6309	78	115.0540
4	1.3802	29	30.9465	54	71.3633	79	116.9516
5	2.0792	30	32.4237	55	73.1037	80	118.8547
6	2.8573	31	33.9150	56	74.8519	81	120.7632
7	3.7024	32	35.4202	57	76.6077	82	122.6770
8	4.6055	33	36.9387	58	78.3712	83	124.5961
9	5.5598	34	38.4702	59	80.1420	84	126.5204
10	6.5598	35	40.0142	60	81.9202	85	128.4498
11	7.6012	36	41.5705	61	83.7055	86	130.3843
12	8.6803	37	43.1387	62	85.4979	87	132.3238
13	9.7943	38	44.7185	63	87.2972	88	134.2683
14	10.9404	39	46.3096	64	89.1034	89	136.2177
15	12.1165	40	47.9117	65	90.9163	90	138.1719
16	13.3206	41	49.5244	66	92.7359	91	140.1310
17	14.5511	42	51.1477	67	94.5620	92	142.0948
18	15.8063	43	52.7812	68	96.3945	93	144.0633
19	17.0851	44	54.4246	69	98.2333	94	146.0364
20	18.3861	45	56.0778	70	100.0784	95	148.0141
21	19.7083	46	57.7406	71	101.9297	96	149.9964
22	21.0508	47	59.4127	72	103.7870	97	151.9831
23	22.4125	48	61.0939	73	105.6503	98	153.9744
24	23.7927	49	62.7841	74	107.5196	99	155.9700
25	25.1907	50	64.4831	75	109.3946	100	157.9700

[a]This table also allows calculation of $\log m = \log\{m!/(m-1)!\} = \log m! - \log(m-1)!$.

Illustration

$$p(7) = \frac{\binom{13}{7}\binom{39}{6}}{\binom{52}{13}}$$

$$= \frac{13!}{7!6!} \times \frac{39!}{6!33!} \times \frac{13!39!}{52!}.$$

We now make use of table 6.1:

log 13! =	9.7943	log 7! =	3.7024
log 39! =	46.3096	log 6! =	2.8573
log 13! =	9.7943	log 6! =	2.8573
log 39! =	46.3096	log 33! =	36.9387
	112.2078	log 52! =	67.9067
	−114.2624 ⟵		114.2624.
	$\overline{3}$.9454		

Now

$$\text{antilog}(\overline{3}.9454) = 10^{-3}\,\text{antilog}(0.9454)$$
$$= 10^{-3}\{8.710 + \tfrac{54}{100}(8.913 - 8.710)\}$$
$$= 10^{-3}\{8.820\}$$
$$= 0.00882.$$

Example. Suppose that a company has 99 employees and that men outnumber women 2 to 1. What is the probability that, in a randomly selected group of 20, half will be men and half will be women?

Here $N = 99$, $n = 20$, $W = 99(\tfrac{2}{3}) = 66$, $N - W = 99(\tfrac{1}{3}) = 33$, and $x = 10$, so that

$$p(10) = \frac{\binom{66}{10}\binom{33}{10}}{\binom{99}{20}} = 0.0455.$$

That is, the chance is only about $4\tfrac{1}{2}\%$ that half the group will be men and half women. (The reader should check this answer as an exercise.)

Table 6.2.   A Short Table of Common Antilogarithms[a]

	0	1	2	3	4	5	6	7	8	9
0.0	1000	1023	1047	1072	1096	1122	1148	1175	1202	1230
0.1	1259	1288	1318	1349	1380	1413	1445	1479	1514	1549
0.2	1585	1622	1660	1698	1738	1778	1820	1862	1905	1950
0.3	1995	2042	2089	2138	2188	2239	2291	2344	2399	2455
0.4	2512	2570	2630	2692	2754	2818	2884	2951	3020	3090
0.5	3162	3236	3311	3388	3467	3548	3631	3715	3802	3890
0.6	3981	4074	4169	4266	4365	4467	4571	4677	4786	4898
0.7	5012	5129	5248	5370	5495	5623	5754	5888	6026	6166
0.8	6310	6457	6607	6761	6918	7079	7244	7413	7586	7762
0.9	7943	8128	8318	8511	8710	8913	9120	9333	9550	9772

[a] Place a decimal point after the first figure in the entry found. Example: The antilogarithm of 2.63 is $10^2 \times (\text{antilog } 0.63) = 10^2 \times 4.266 = 426.6$.

Exercises

1. If an urn contains 40 white and 20 red balls, and if 10 balls are drawn from it without replacement, find the probability that 5 are white.
2. Find the probability that a hand of 13 cards drawn from a bridge deck will contain 8 spades.
3. A committee of 5 people is to be selected at random from a group of 12 men and 12 women. Find the probability that the committee will
   (a) have no women members.
   (b) be either "all male" or "all female."
4. One-fifth of the 30 students in a particular class are math majors. A group of 6 students is chosen at random. Find the probability that the group will have
   (a) 2 math majors.   (b) no math majors.
5. Find the probability that a poker hand (5 cards) will contain
   (a) 4 aces.   (b) 1 ace.   (c) no ace.
6. (a) What is the probability that a bridge hand will contain no spades or diamonds?
   (b) What is the probability that a bridge hand will contain only two suits?
7. A room has three empty lamp sockets. From six good and four defective lamp bulbs, three are selected at random. What is the probability that some light will appear in the room when the chosen bulbs are inserted?
8. Find the probability that a poker hand of five cards will be valued at "three of a kind" (exactly three cards of the same value, and two other cards, not of the same value). Assume that the five cards are dealt at random from an ordinary deck.
9. Find the probability that in a hand of five cards drawn at random from an ordinary deck, there will be a pair of kings, another pair of different value, and a fifth card not having the same value as any of the others.

   THE DISTRIBUTION FUNCTION.   For the hypergeometric distribution, the distribution function is

$$F(x) = \frac{\sum_{t \le x} \binom{W}{t} \binom{N - W}{n - t}}{\binom{N}{n}}.$$

$F(x)$ is the probability that there will be x or fewer successes.

Example. The probability that a hand of 13 cards from a bridge deck will contain two or fewer spades is

$$P(X \leq 2) = F(2) = \frac{\sum_{t=0}^{2} \binom{13}{t}\binom{52-13}{13-t}}{\binom{52}{13}}$$

$$= \frac{\binom{13}{0}\binom{39}{13} + \binom{13}{1}\binom{39}{12} + \binom{13}{2}\binom{39}{11}}{\binom{52}{13}}$$

$$= 0.01279 + 0.08008 + 0.2059$$

$$= 0.29877$$

$$= 0.30 \qquad \text{to two decimal places.}$$

In other words, there is about a 30% chance that the hand will contain two or fewer spades.

Exercises

1. If an urn contains 40 white and 20 red balls, and if 10 balls are drawn from it without replacement, find the probability that 3 or fewer are white.
2. Find the probability that a hand of 13 cards drawn from a bridge deck will contain fewer than two spades.
3. One-fifth of the 30 students in a particular class are math majors. A group of 6 students is chosen at random. Find the probability that the group will have at least 2 math majors. [*Hint*: $1 - P$(group will have fewer than 2 math majors).]
4. Find the probability that a poker hand (5 cards) will contain at least 2 aces. [*Hint*: $1 - P$(hand contains fewer than 2 aces).]
5. A man eats 5 cookies from a package containing 20 cookies, 6 of which are spoiled. If he eats 3 or more spoiled cookies, he will get sick. Find the probability that he will get sick.

AN IDENTITY. For a hypergeometric distribution we must have

$$F(W) = \sum_{x=0}^{W} p(x) = 1,$$

because the $p(x)$ represent all the possibilities that can occur; it follows that

$$\sum_{x=0}^{W} \binom{W}{x}\binom{N-W}{n-x} = \binom{N}{n}.$$

We shall make use of this in evaluating the mean and variance of the hypergeometric distribution.

MEAN OF A HYPERGEOMETRIC DISTRIBUTION. We see that

$$\binom{N}{n} EX = \sum_{x=0}^{W} x \binom{W}{x} \binom{N-W}{n-x}$$

$$= W \sum_{x=1}^{W} \frac{(W-1)!}{(x-1)!(W-x)!} \binom{N-W}{n-x}$$

(canceling out x in the denominator and removing a factor W)

$$= W \sum_{s=0}^{W-1} \frac{(W-1)!}{s!(W-s-1)!} \binom{N-W}{n-s-1}$$

(setting $s = x - 1$, that is, $x = s + 1$, so that, as x goes from 1 to W, s goes from 0 to $W - 1$)

$$= W \sum_{s=0}^{W-1} \binom{W-1}{s} \binom{N-1-(W-1)}{n-1-s}$$

$$= W \binom{N-1}{n-1},$$

using the identity given above. It follows that

$$EX = W \binom{N-1}{n-1} \bigg/ \binom{N}{n}$$

$$= W \times \frac{(N-1)!}{(n-1)!(N-n)!} \times \frac{n!(N-n)!}{N!}$$

$$= n \frac{W}{N}.$$

(*Note*: If we were sampling *with* replacement, $p = W/N$ would be the appropriate binomial parameter and the binomial mean would be $np = nW/N$, the same as for the hypergeometric distribution.)

VARIANCE OF A HYPERGEOMETRIC DISTRIBUTION. We use here the same trick as we used for the binomial distribution; that is, we first find $EX(X - 1)$. The process is as follows:

$$\binom{N}{n} EX(X-1) = \sum_{x=0}^{W} x(x-1) \binom{W}{x} \binom{N-W}{n-x}$$

$$= W(W-1) \sum_{x=2}^{W} \frac{(W-2)!}{(x-2)!(W-x)!} \binom{N-W}{n-x}.$$

(Now set $x = s + 2$.)

$$= W(W - 1) \sum_{s=0}^{W-2} \frac{(W - 2)!}{s!(W - s - 2)!} \binom{N - W}{n - 2 - s}$$

$$= W(W - 1) \sum_{s=0}^{W-2} \binom{W - 2}{s} \binom{N - 2 - (W - 2)}{n - 2 - s}.$$

(Use the identity.)

$$= W(W - 1) \binom{N - 2}{n - 2}.$$

Thus

$$EX(X - 1) = W(W - 1) \binom{N - 2}{n - 2} \Big/ \binom{N}{n}$$

$$= W(W - 1) \frac{n(n - 1)}{N(N - 1)}.$$

It follows that

$$V(X) = EX(X - 1) + EX - (EX)^2$$

$$= W(W - 1) \frac{n(n - 1)}{N(N - 1)} + n \frac{W}{N} - n^2 \frac{W^2}{N^2}$$

$$= n \frac{W}{N} \left(1 - \frac{W}{N} \right) \left(\frac{N - n}{N - 1} \right)$$

after some reduction.

Note. If we were sampling *with* replacement, $p = W/N$ would be the appropriate binomial parameter and the binomial variance would be

$$np(1 - p) = n \frac{W}{N} \left(1 - \frac{W}{N} \right).$$

This is slightly greater than the hypergeometric variance because of the factor $(N - n)/(N - 1)$ in the latter. Note that, if N is very large compared with n, this factor is close to one. This makes sense, as the larger the hypergeometric population is, the less will the probability of a success change from trial to trial, in other words, the more nearly will X behave like a binomial variable rather than a hypergeometric. Another way of saying this loosely is that, as N becomes very large compared with n (the number of trials), the hypergeometric distribution tends to the binomial distribution.

Exercises

1. An urn contains 40 white and 20 red balls, and 10 balls are drawn from it without replacement. If X = the number of white balls drawn, find EX and $V(X)$.
2. A hand of 13 cards is drawn from a bridge deck, and X = the number of spades in the hand. Find EX and $V(X)$.
3. One-fifth of the 30 students in a particular class are math majors. A group of 6 students is chosen at random and contains X math majors. Find EX and $V(X)$.
4. Let X = the number of aces in a poker hand of 5 cards. Find EX and $V(X)$.
5. A man eats 5 cookies from a package containing 20 cookies, 6 of which are spoiled. Let X = the number of spoiled cookies he eats. Find EX and $V(X)$.

MOMENTS OF HYPERGEOMETRIC DISTRIBUTION. We recall that moments of a distribution were defined in Section 5.3. By working out

$$EX(X - 1)(X - 2) \quad \text{and} \quad EX(X - 1)(X - 2)(X - 3)$$

we can show, after some manipulation, that

$$\mu_3 = \frac{npq(q - p)(N - n)(N - 2n)}{(N - 1)(N - 2)},$$

$$\mu_4 = \frac{npq(N - n)}{(N - 1)(N - 2)(N - 3)}Q,$$

where

$$Q = N(N + 1) - 6n(N - n) + 3pq\{N^2(n - 2) - Nn^2 + 6n(N - n)\},$$
$$p = \frac{W}{N},$$

and

$$q = 1 - p = 1 - \frac{W}{N}.$$

In the same notation we note that

$$\mu_2 = V(X) = \frac{npq(N - n)}{N - 1}.$$

Again note that, if N is very large compared with n, μ_3 and μ_4, as well as μ_2, tend to the corresponding binomial moments (given in Exercises 26 and 27 on page 121).

Exercise

1. Verify the expressions μ_3 and μ_4 above for the hypergeometric distribution.

6.2. Poisson Distribution

A random variable closely related to the binomial variable is one whose possible values $0, 1, 2, \ldots$ represent the number of occurrences of some outcome, not in a given number of trials, but in a given period of time or region of space. This is a Poisson variable, defined as follows.

Definition. If X is a random variable with possible values $0, 1, 2, \ldots$, and if $P(X = x)$ is given by

$$p(x) = \frac{e^{-\lambda}\lambda^x}{x!} \qquad x = 0, 1, 2, \ldots,$$

then we say that X has a *Poisson distribution.* The Greek letter λ (lambda) is called the parameter of the distribution. (Note that there is only one parameter in this distribution; the binomial distribution has two.) The letter e denotes a well-known constant, value $e = 2.71828 \ldots$. Tables of e to various powers exist and e is the base of the so-called natural logarithms.

The Poisson distribution is useful as a probability model in its own right for many practical random phenomena and can also be used as an approximation to the binomial distribution for certain ranges of values of the binomial parameters n and p. We now look at some examples of Poisson distributions.

Example 1. Suppose $\lambda = 1$. Then we obtain the following to two decimal places:

x	0	1	2	3	4	5 up
$p(x)$	0.37	0.37	0.18	0.06	0.02	—

This distribution is shown in Figure 6.1(a).

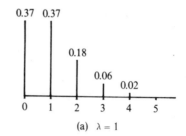

(a) $\lambda = 1$

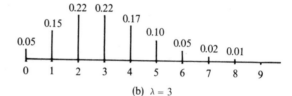

(b) $\lambda = 3$

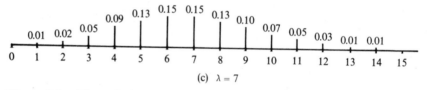

(c) $\lambda = 7$

Figure 6.1. Three Poisson distributions with different parameter values:
(a) $\lambda = 1$; (b) $\lambda = 3$; (c) $\lambda = 7$.

Example 2. Suppose $\lambda = 3$. Then we obtain, to two decimal places:

x	0	1	2	3	4	5	6	7	8	9 up
$p(x)$	0.05	0.15	0.22	0.22	0.17	0.10	0.05	0.02	0.01	—

This distribution is shown in Figure 6.1(b).

Example 3. Suppose $\lambda = 7$. Then we obtain, to two decimal places:

x	0	1	2	3	4	5	6	7
$p(x)$	—	0.01	0.02	0.05	0.09	0.13	0.15	0.15
x	8	9	10	11	12	13	14	15 up
$p(x)$	0.13	0.10	0.07	0.05	0.03	0.01	0.01	—

This distribution is shown in Figure 6.1(c).

Comment. We see that, as λ increases, the peak of the distribution moves more and more to the right, and the distribution tends to become more and more symmetrical. For very small λ values, the peak of the distribution is at $x = 0$ and the distribution falls off rapidly as x increases. Table 6.3 gives a short table of individual Poisson probabilities for a selection of values of λ, to three decimal places. (For a more complete table see, for example, *Handbook of Tables for Probability and Statistics*, W. H. Beyer (ed.), Chemical Rubber Co., Cleveland, 2nd ed., 1968.)

Note that, in Table 6.3, all entries in a column for a specified value of λ should add to one. Where they do not, it is due to the fact that rounding to three figures has induced small errors that have accumulated rather than canceled out. Values of $p(x)$ for other values of λ can be worked out quite easily using a table of exponentials $e^{-\lambda}$.

Table 6.3. Individual Terms of the Poisson Distribution for Certain Values of λ [a]

			λ		
x	0.2	0.4	0.6	0.8	1.0
0	.819	.670	.549	.449	.368
1	.164	.268	.329	.360	.368
2	.016	.054	.099	.144	.184
3	.001	.007	.020	.038	.061
4	—	.001	.003	.008	.015
5	—	—	—	.001	.003
6	—	—	—	—	.001

			λ		
x	1.2	1.4	1.6	1.8	2.0
0	.301	.247	.202	.165	.135
1	.361	.345	.323	.298	.271
2	.217	.242	.258	.268	.271
3	.087	.113	.138	.161	.180
4	.026	.040	.055	.072	.090
5	.006	.011	.018	.026	.036
6	.001	.003	.005	.008	.012
7	—	.001	.001	.002	.003
8	—	—	—	.001	.001

Table 6.3—*continued*

			λ			
x	2.5	3.0	3.5	4.0	4.5	5.0
0	.082	.050	.030	.018	.011	.007
1	.205	.149	.106	.073	.050	.034
2	.257	.224	.185	.147	.113	.084
3	.214	.224	.216	.195	.169	.140
4	.134	.168	.189	.195	.19Q	.176
5	.067	.101	.132	.156	.171	.176
6	.028	.050	.077	.104	.128	.146
7	.010	.022	.039	.060	.082	.104
8	.003	.008	.017	.030	.046	.065
9	.001	.003	.007	.013	.023	.036
10	—	.001	.002	.005	.010	.018
11	—	—	.001	.002	.004	.008
12	—	—	—	.001	.002	.003
13	—	—	—	—	.001	.001
14	—	—	—	—	—	.001

			λ		
x	6	7	8	9	10
0	.003	.001	—	—	—
1	.015	.006	.003	.001	.001
2	.045	.022	.011	.005	.002
3	.089	.052	.029	.015	.008
4	.134	.091	.057	.034	.019
5	.161	.128	.092	.061	.038
6	.161	.149	.122	.091	.063
7	.138	.149	.140	.117	.090
8	.103	.130	.140	.132	.113
9	.069	.101	.124	.132	.125
10	.041	.071	.099	.119	.125
11	.023	.045	.072	.097	.114
12	.011	.026	.048	.073	.095
13	.005	.014	.030	.050	.073
14	.002	.007	.017	.032	.052

Table 6.3—*continued*

			λ		
x	6	7	8	9	10
15	.001	.003	.009	.019	.035
16	—	.001	.005	.011	.022
17	—	.001	.002	.006	.013
18	—	—	.001	.003	.007
19	—	—	—	.001	.004
20	—	—	—	.001	.002
21	—	—	—	—	.001

[a] Missing entries or entries not listed (e.g., $\lambda = 1.0$, $x = 7$) are zero to three decimal places.

Exercises

1. Given that $e^{-3} = 0.050$ to three decimal places, obtain the $\lambda = 3$ column in Table 6.3.
2. Given that $e^{-0.7} = 0.500$ to three decimal places, obtain a column for $\lambda = 0.7$.
3. Given that $e^{-0.4} = 0.670$ to three decimal places, obtain the $\lambda = 0.4$ column in Table 6.3.

AN IDENTITY. The following result is needed to sum all the terms of a Poisson distribution:

$$e^{\lambda} = \sum_{x=0}^{\infty} \frac{\lambda^x}{x!}$$

$$= 1 + \lambda + \frac{\lambda^2}{2!} + \frac{\lambda^3}{3!} + \frac{\lambda^4}{4!} + \cdots + \frac{\lambda^x}{x!} + \cdots.$$

We do not prove this result because the calculus methods needed are beyond the scope of this book. We ask the reader to accept it without proof. Using the result, however, we see that, for a Poisson distribution,

$$\sum_{x=0}^{\infty} p(x) = \sum_{x=0}^{\infty} \frac{e^{-\lambda}\lambda^x}{x!} = e^{-\lambda} \sum_{x=0}^{\infty} \frac{\lambda^x}{x!}$$

$$= e^{-\lambda}e^{\lambda}$$

$$= 1.$$

THE DISTRIBUTION FUNCTION. For a Poisson distribution the distribution function is

$$F(x) = \sum_{t=0}^{x} \frac{e^{-\lambda}\lambda^t}{t!}.$$

The identity above implies that $F(\infty) = 1$, of course.

Example. If the number of telephone calls per minute entering a certain switchboard follows a Poisson probability law with $\lambda = 5$, the probability that there will be 2 or fewer calls in any given minute is

$$P(X \le 2) = F(2)$$

$$= \sum_{t=0}^{2} \frac{e^{-5}5^t}{t!}$$

$$= e^{-5}\left\{\frac{5^0}{0!} + \frac{5^1}{1!} + \frac{5^2}{2!}\right\}$$

$$= e^{-5}(18.5)$$

$$= 0.125,$$

since $e^{-5} = 0.006738$. Of course we could also have obtained this result from Table 6.3, adding together the top three entries in the $\lambda = 5$ column.

MEAN OF A POISSON DISTRIBUTION. We see that

$$EX = \sum_{x=0}^{\infty} x\frac{e^{-\lambda}\lambda^x}{x!}$$

$$= \sum_{x=1}^{\infty} \frac{e^{-\lambda}\lambda^x}{(x-1)!}$$

$$= \lambda \sum_{x=1}^{\infty} \frac{e^{-\lambda}\lambda^{x-1}}{(x-1)!}.$$

If we now set $s = x - 1$, then as x goes from 1 to ∞, s goes from 0 to ∞, and we get

$$EX = \lambda \sum_{s=0}^{\infty} \frac{e^{-\lambda}\lambda^s}{s!} = \lambda,$$

since the summation is equal to one by the identity above. In other words, λ is the mean, or average number of occurrences, of a phenomenon which follows a Poisson distribution with parameter λ.

VARIANCE OF A POISSON DISTRIBUTION. We need

$$V(X) = EX^2 - (EX)^2$$

$$= EX(X - 1) + EX - (EX)^2,$$

using the same manipulation as for the binomial and hypergeometric distribution cases. Now

$$EX(X - 1) = \sum_{x=0}^{\infty} x(x - 1)\frac{e^{-\lambda}\lambda^x}{x!}$$

$$= \lambda^2 \sum_{x=2}^{\infty} \frac{e^{-\lambda}\lambda^{x-2}}{(x - 2)!}$$

$$= \lambda^2,$$

again using the summation identity given earlier. Since $EX = \lambda$, it follows that, for the Poisson distribution

$$V(X) = \lambda^2 + \lambda - \lambda^2$$

$$= \lambda.$$

In other words, the mean *and the variance* of a Poisson distribution are identical and are both equal to λ, the parameter of the distribution.

TWO EXAMPLES OF APPLICATIONS OF THE POISSON DISTRIBUTION.

As we mentioned earlier in this section, the Poisson distribution is useful as a model for various common types of random phenomena, particularly when X is the number of occurrences of an outcome in (1) an interval of time, or (2) an interval of space.

An example of the first type is this: Suppose X is the number of radioactive particles emitted in 1 minute by a given mineral. Then λ is the average number of particles emitted in 1 minute, and the probability that x particles will be emitted in 1 minute is given by $P(X = x) = p(x) = e^{-\lambda}\lambda^x/x!$. The probability that x particles will be emitted in a 2-minute interval is given by $P(X = x) = p(x) = e^{-2\lambda}(2\lambda)^x/x!$, for now the average number of particles emitted in the given interval of 2 minutes is 2λ. (The point to remember is that the parameter of the Poisson distribution is the average or mean value of X for the given time interval.)

An example of the second type is this: Suppose X is the number of dandelions per square yard in a lawn. Let λ be the average number of dandelions present per square yard. The probability that, in a given area of 1 square yard, there will be x dandelions, is given by $P(X = x) = p(x) = e^{-\lambda}\lambda^x/x!$. For example, if the average number of dandelions in a square yard is $\lambda = 3$, then the probability that a particular 1-square-yard plot will have five dandelions is given by $p(5) = e^{-3}3^5/5! = 0.101$; the probability that there will be fewer than two dandelions is $P(X < 2) = \sum_{x=0}^{1} e^{-3}3^x/x! = 0.199$. The probability that there will be one dandelion *in a 1-square-foot area* is $e^{-1/3}(\frac{1}{3})/1! = 0.239$, because now the average per square foot is $(\frac{1}{9}) \times 3 = \frac{1}{3}$.

Exercises

1. If the number of telephone calls on a certain line in 1 hour is a Poisson variable with $\lambda = 10$, find the probability that
   (a) there will be exactly 10 calls on this line in a given hour.
   (b) there will be more than 15 calls in an hour.
   (c) there will be 20 calls in a 2-hour period.
   (d) two consecutive 1-hour periods will each have 10 calls.
2. The average number of errors per page of certain typeset material is 1.8. Find the probability that there will be, on a single page,
   (a) one error.
   (b) more than one error.
   (c) fewer than five errors.
3. A certain typist makes an average of one error in 400 words. Use a Poisson distribution to find the probability that this typist will make, in a 1000-word letter,
   (a) 3 errors.   (c) no errors.
   (b) 5 errors.   (d) at least 3 errors.
4. A salesman averages one sale for every 50 doorbells he rings. In a given day he rings 100 doorbells. Using a Poisson distribution, find the probability that he will make
   (a) 5 sales.   (b) no sale.   (c) at least one sale.
5. The number of auto accidents at a busy intersection averages 2 per 10,000 automobiles. During a particular week 40,000 automobiles entered the intersection. Using a Poisson distribution, find the probability of
   (a) no accidents.   (b) 5 accidents.   (c) at least one accident.
6. The number of oak trees in a certain national forest has a Poisson distribution with $\lambda = 25$ trees per acre. Find the probability that a particular one-acre plot in this forest will have 30 oak trees. You need not work this out to an answer; simply give the formula with the proper values.
7. Suppose we know that X has a Poisson distribution with mean $\lambda = 1$. Find Chebyshev's bound, and also the correct value, for the probability that an observed value of X lies within $2\frac{1}{2}$ standard deviations of the mean. Comment on the comparison.

THE POISSON AS AN APPROXIMATION TO THE BINOMIAL DISTRIBUTION.

Let n and p be the parameters of a binomial distribution, whose mean is np and whose variance is $np(1 - p)$. Suppose n becomes very large (that is, "tends to infinity") and p becomes very small (that is, "tends to zero") simultaneously, in such a way that the product $\lambda = np$ remains fixed. In other words, we set $p = \lambda/n$, where λ is a fixed value, and let n increase. Then as n increases, the binomial probabilities

$$p(x) = \binom{n}{x} p^x (1 - p)^{n-x} \qquad x = 0, 1, 2, \ldots, n$$

get closer and closer ("tend to") the Poisson probabilities

$$p(x) = \frac{e^{-\lambda} \lambda^x}{x!} \qquad \text{where } \lambda = np, x = 0, 1, 2, \ldots.$$

The proof of this involves ideas beyond the scope of this book, and we ask the reader to accept the result without proof.

The point of this result is that, when n is "large" and p is "small," we can make use of a Poisson distribution to approximate to binomial probabilities and to cumulative probabilities. This works fairly well for even quite modest values of n, as we see in the example below. The larger is n and the smaller is p, the better the approximation will be.

Example. Suppose $p = 0.1$ and $n = 20$. Then if we set $\lambda = np = 2$, we can use the $\lambda = 2$ column of Table 6.3 to approximate the corresponding values of the binomial probabilities. How good is this approximation in this case? The comparison in Table 6.4 is given to three decimal places.

Table 6.4. Comparison of Binomial Probabilities and Their Poisson Approximations for $p = 0.1$, $n = 20$

x	$p(x)$, binomial	$p(x)$, Poisson, $\lambda = 2$
0	0.122	0.135
1	0.270	0.271
2	0.285	0.271
3	0.190	0.180
4	0.090	0.090
5	0.032	0.036
6	0.009	0.012
7	0.002	0.003
8	—	0.001
9 and higher	—	—

We emphasize again that this is a case where n is really quite small and that, in cases where n is larger and p smaller, much closer agreement would be found.

Example. Suppose X is a random variable with a binomial distribution and the probability of success in a single trial is $p = 0.02$. Then the probability of 5 successes in 100 trials is given exactly by the binomial probability

$$p(5) = \binom{100}{5}(0.02)^5(0.98)^{95}.$$

Although tables of logarithms of numbers and even logarithms of factorials are available, the problem still requires a fair amount of calculation. However, if we use the Poisson approximation to the binomial with $\lambda = np = 100(0.02) = 2$, we obtain the approximate probability

$$p(5) = e^{-2}2^5/5! = 0.036$$

from Table 6.3. The exact probability, obtained by using the binomial distribution, is 0.035, in fact.

Exercises

1. If a machine produces 2% defective items, find the probability that, of 100 items taken from the day's production, exactly 5 will be defective. First use the binomial distribution and then approximate the probability using a Poisson distribution.

2. In the framework of Exercise 1, what is the probability that 4 or fewer of the 100 items will be defective? (Again, do both calculations and compare.)

3. On the average, only one student in a hundred achieves a perfect paper on a certain French examination. Use a Poisson distribution as an approximation to the binomial to obtain the probability that three or more students in a class of 200 will achieve perfection. ($e^{-0.02} = 0.98$.)

4. State Chebyshev's inequality when $h = 2$. Now suppose that X has a binomial distribution with $p = 1/37$ and $n = 111$. Use the Poisson approximation to the binomial to approximate the true probability and compare it with the Chebyshev result.

CHAPTER 7

Joint Probability Functions, Regression Curves, Bivariate Moments and Correlation, and the Multinomial Distribution

7.1. Joint Probability Functions

Up to this point we have considered the distribution of just *one* characteristic of the outcome of a random phenomenon. In many situations we are interested in observing two or more characteristics simultaneously and finding what the *joint* probability distribution of these characteristics may be. For example, when drawing a card from a deck we may be interested not just in the suit *or* the numerical value, but in *both* simultaneously; we may be interested in both the height *and* weight of a person; we may want to observe both the make and the age of an automobile passing a certain check point.

TWO-DIMENSIONAL RANDOM VARIABLES. In the notation of random variables we write the observed values of two numerically valued characteristics, X and Y, of a random phenomenon as an ordered number pair (x, y). The random variables X and Y are numerically valued functions defined on a joint sample space; they may each be discrete or continuous. We shall consider only the case in which both X and Y are discrete random variables. We shall call the pair (X, Y) a *two-dimensional* or *bivariate random variable*.

Definition. A joint (or *bivariate*) *probability function*, $p(x, y)$, of a two-dimensional discrete random variable (X, Y) is a function

$$p(x, y) = P(X = x \text{ and } Y = y),$$

which satisfies the two conditions:

1. $p(x, y) \geq 0$ for all x and y.
2. $\sum_x \sum_y p(x, y) = 1$, the summation being over all possible values of x and y.

This is the obvious extension of the definition of the probability function for a *single* discrete random variable.

Example. Suppose an urn contains six balls marked with the number 1, five balls marked with 2, and four balls marked with 3; one ball is drawn and then, without replacement of the first, a second ball is drawn. Let X represent the number on the first ball drawn and Y the number on the second ball drawn. Thus (X, Y) is a two-dimensional random variable with possible values $(1, 1), (1, 2), (1, 3), (2, 1), (2, 2), (2, 3), (3, 1), (3, 2), (3, 3)$. This then is the total sample space. The probabilities attached to the various sample points are assigned in accordance with previous work; for example,

$$P(1, 1) = P(X = 1 \text{ and } Y = 1) = P(X = 1)P(Y = 1 | X = 1) = \left(\tfrac{6}{15}\right)\left(\tfrac{5}{14}\right) = \tfrac{1}{7}.$$

The nine probabilities can be conveniently arranged as in Table 7.1.

Table 7.1. A Bivariate (or Joint) Probability Function for Two Variables X and Y

	y			Row sums
x	1	2	3	(marginal probabilities)
1	$\frac{1}{7}$	$\frac{1}{7}$	$\frac{4}{35}$	$\frac{2}{5}$
2	$\frac{1}{7}$	$\frac{2}{21}$	$\frac{2}{21}$	$\frac{1}{3}$
3	$\frac{4}{35}$	$\frac{2}{21}$	$\frac{2}{35}$	$\frac{4}{15}$
Column sums (marginal probabilities)	$\frac{2}{5}$	$\frac{1}{3}$	$\frac{4}{15}$	1

DIAGRAMS FOR JOINT PROBABILITY FUNCTIONS.   It is sometimes convenient to draw a diagram so that joint probability functions can
be better visualized. In such a diagram, x and y are plotted on a two-
dimensional base, and the values $p(x, y)$ are drawn upward at (x, y) out from
this base, the heights being proportional to the value of $p(x, y)$. A representation of the distribution in Table 7.1 is shown in Figure 7.1. Note that there
are many possible variations of such a figure. Solid (square) blocks can
replace "poles" and the axes can be oriented in various ways. Our figure is
drawn in an orientation similar to that of our table. If the x and y axes are
oriented differently in a figure, the table can be rearranged to correspond
to the particular orientation selected. The reader may find it helpful to
construct similar figures for the various joint probability distributions he
meets in this book.

MARGINAL PROBABILITIES.   The probabilities in the "sum" row
and "sum" column of Table 7.1 are aptly called *marginal probabilities*; they
give the probability of a particular value of one of the variables irrespective
(that is, without specifying) the value of the other variable.

Example.   Using the example above,

$$P(X = 1) = P[(X = 1) \text{ and } (Y = 1 \text{ or } Y = 2 \text{ or } Y = 3)]$$
$$= \tfrac{1}{7} + \tfrac{1}{7} + \tfrac{4}{35} = \tfrac{14}{35} = \tfrac{2}{5},$$
$$P(Y = 2) = P[(Y = 2) \text{ and } (X = 1 \text{ or } X = 2 \text{ or } X = 3)]$$
$$= \tfrac{1}{7} + \tfrac{2}{21} + \tfrac{2}{21} = \tfrac{7}{21} = \tfrac{1}{3}.$$

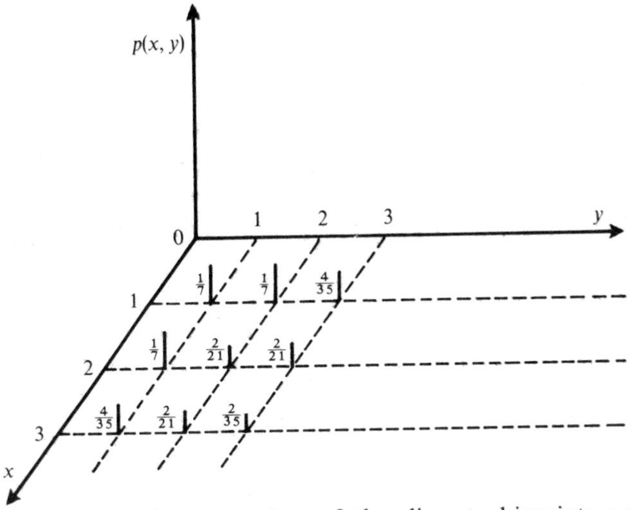

Figure 7.1.   Pictorial representation of the discrete bivariate probability
function given in Table 7.1.

DEFINITION OF MARGINAL PROBABILITIES. Formally we define the marginal probabilities in general as

$$p(x) = P(X = x) = \sum_y p(x, y)$$

and

$$q(y) = P(Y = y) = \sum_x p(x, y).$$

(We must use another letter, q, here or we shall have confusion in numerical examples, as we shall appreciate in a moment.) Note that, in the two-way table in the example,

$$\sum_x p(x) = \sum_y q(y) = 1.$$

This is true generally, because each is equivalent to

$$\sum_x \sum_y p(x, y)$$

by definition.

CONDITIONAL PROBABILITY. The definition of conditional probability for random variables is similar to that for events. In general, the conditional probability of $X = x$ given $Y = y$ is

$$p(x|y) = \frac{p(x, y)}{q(y)}, \qquad \text{provided } q(y) \neq 0,$$

and the conditional probability that $Y = y$ given that $X = x$ is

$$q(y|x) = \frac{p(x, y)}{p(x)}, \qquad \text{provided } p(x) \neq 0.$$

Example. In the main example above,

$$P(X = 2| Y = 3) = p(2|3) = \frac{p(2, 3)}{q(3)} = \frac{\frac{2}{21}}{\frac{4}{15}} = \frac{5}{14},$$

$$P(Y = 3|X = 1) = q(3|1) = \frac{p(1, 3)}{p(1)} = \frac{\frac{4}{35}}{\frac{2}{5}} = \frac{2}{7}.$$

INDEPENDENCE OF TWO RANDOM VARIABLES. Independence of two random variables is defined in a manner similar to that for independence of two events. Two random variables X and Y are *independent* if and only if

$$p(x, y) = p(x)q(y),$$

that is, if and only if the joint probability function is the product of the two marginal probability functions for every x and y.

Example.   If the joint probabilities for X and Y are as given in Table 7.2 and as illustrated in Figure 7.2, it can be verified that X and Y are independent.

Table 7.2.   A Bivariate Probability Distribution

	y			Row Sums [marginal probabilities $p(x)$]
x	1	2	3	
1	$\frac{1}{24}$	$\frac{1}{16}$	$\frac{1}{48}$	$\frac{1}{8}$
2	$\frac{1}{12}$	$\frac{1}{8}$	$\frac{1}{24}$	$\frac{1}{4}$
3	$\frac{5}{24}$	$\frac{5}{16}$	$\frac{5}{48}$	$\frac{5}{8}$
Column sums [marginal probabilities $q(y)$]	$\frac{1}{3}$	$\frac{1}{2}$	$\frac{1}{6}$	1

The marginal probabilities are just the sums of rows, for $p(x)$, and the sums of columns, for $q(y)$. We find that

$$p(1) = \tfrac{1}{8},\ p(2) = \tfrac{1}{4},\ p(3) = \tfrac{5}{8},$$
$$q(1) = \tfrac{1}{3},\ q(2) = \tfrac{1}{2},\ q(3) = \tfrac{1}{6}.$$

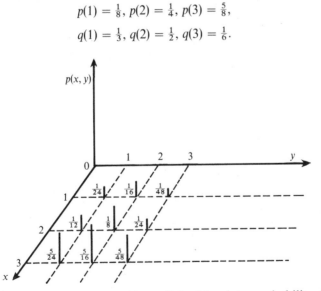

Figure 7.2.   Pictorial representation of the bivariate probability function given in Table 7.2.

The joint probability function is the product of the marginal functions for every value of x and y. For example,

$$p(1, 1) = \tfrac{1}{24} = (\tfrac{1}{8})(\tfrac{1}{3}) = p(1)q(1),$$

$$p(1, 2) = \tfrac{1}{16} = (\tfrac{1}{8})(\tfrac{1}{2}) = p(1)q(2),$$

$$p(3, 3) = \tfrac{5}{48} = (\tfrac{5}{8})(\tfrac{1}{6}) = p(3)q(3),$$

and similarly for the other (x, y) values. (The reader should check the other cases for himself.) It follows that X and Y are independent.

THE DISTRIBUTION FUNCTION. The corresponding joint distribution function is given by

$$F(x, y) = P(X \leq x \text{ and } Y \leq y) = \sum_{u \leq x} \sum_{v \leq y} p(u, v).$$

[Note the difference between the *distribution* (which consists of all the values which the variables (X, Y) can take and the associated probabilities) and the *distribution function* (which is the specific function of x and y given above).]

Example. The distribution function $F(x, y)$ for the joint distribution given in Table 7.2 is as follows:

$$F(x, y) = 0, \qquad\qquad\qquad \text{if } x < 1 \text{ or } y < 1,$$

$$F(x, y) = p(1, 1) = \tfrac{1}{24}, \qquad\qquad \text{if } 1 \leq x < 2, \quad 1 \leq y < 2,$$

$$F(x, y) = p(1, 1) + p(1, 2)$$
$$= \tfrac{5}{48}, \qquad\qquad\qquad \text{if } 1 \leq x < 2, \quad 2 \leq y < 3,$$

$$F(x, y) = p(1, 1) + p(1, 2) + p(1, 3)$$
$$= \tfrac{1}{8}, \qquad\qquad\qquad \text{if } 1 \leq x < 2, \quad 3 \leq y,$$

$$F(x, y) = p(1, 1) + p(2, 1) = \tfrac{1}{8}, \qquad \text{if } 2 \leq x < 3, \quad 1 \leq y < 2,$$

$$F(x, y) = p(1, 1) + p(2, 1) + p(3, 1)$$
$$= \tfrac{1}{3}, \qquad\qquad\qquad \text{if } 3 \leq x, \quad 1 \leq y < 2,$$

$$F(x, y) = p(1, 1) + p(1, 2)$$
$$+ p(2, 1) + p(2, 2)$$
$$= \tfrac{5}{16}, \qquad\qquad\qquad \text{if } 2 \leq x < 3, \quad 2 \leq y < 3,$$

$$F(x, y) = p(1, 1) + p(1, 2) + p(1, 3)$$
$$+ p(2, 1) + p(2, 2) + p(2, 3)$$
$$= \tfrac{3}{8}, \qquad\qquad\qquad \text{if } 2 \leq x < 3, \quad 3 \leq y,$$

$$F(x, y) = p(1, 1) + p(1, 2)$$
$$+ p(2, 1) + p(2, 2)$$
$$+ p(3, 1) + p(3, 2)$$
$$= \tfrac{5}{6}, \qquad \text{if } 3 \le x, \quad 2 \le y < 3,$$
$$F(x, y) = 1, \qquad \text{if } 3 \le x \text{ and } 3 \le y.$$

DIAGRAMS FOR JOINT DISTRIBUTION FUNCTIONS. The graph of a bivariate distribution function can also be drawn if desired. The graph consists of a series of planes rising from zero to one in a succession of steps. The levels of these steps for the foregoing example are shown in Figure 7.3.

Exercises

1. In the example of Table 7.1, are X and Y independent random variables?
2. For the distribution in Table 7.1, find the conditional probabilities
   (a) $p(x|1)$ for $x = 1, 2, 3$. What is the sum $\sum_x p(x|1)$?
   (b) $q(y|2)$ for $y = 1, 2, 3$. What is the sum $\sum_y q(y|2)$?
   (c) Do $p(x|y)$ and $q(y|x)$ satisfy the conditions for a probability function? (See Section 4.3.)
3. Make up a joint distribution function of two random variables in which
   (a) the random variables are independent.
   (b) the random variables are *not* independent.

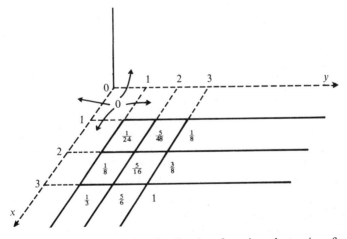

Figure 7.3. Values taken by the distribution function that arises from the probability function given in Table 7.2.

4. A joint probability function is given by the following table:

		y		
x	1	2	3	4
1	$\frac{1}{9}$	$\frac{1}{6}$	$\frac{1}{18}$	$\frac{1}{9}$
2	$\frac{1}{18}$	$\frac{1}{9}$	$\frac{1}{18}$	$\frac{1}{3}$

(a) Draw a figure that shows the joint distribution.
(b) Find the marginal and conditional probabilities.
(c) Are X and Y independent?

5. If $F(x, y)$ is the distribution function obtained from the probability function given in Exercise 4, find

(a) $F(0, 0)$.   (e) $F(1, 3)$.
(b) $F(1, 1)$.   (f) $F(1.7, 3.842)$.
(c) $F(2, 2)$.   (g) $F(2, 4)$.
(d) $F(2.3, 2.9)$.   (h) $F(3, 5)$.

(A diagram like Figure 7.3 may be helpful.)

6. Consider the following bivariate distribution for (X, Y):

		y	
x	0	1	2
0	$\frac{1}{6}$	$\frac{1}{6}$	$\frac{1}{6}$
1	$\frac{1}{12}$	$\frac{1}{3}$	$\frac{1}{12}$

(a) Draw a figure that shows the joint distribution.
(b) Find the marginal and conditional probabilities.
(c) Are X and Y independent?

7. If $F(x, y)$ is the distribution function obtained from the probability function given in Exercise 6, find

(a) $F(-3, -2)$.   (e) $F(0.7, 1.4)$.
(b) $F(0, 0)$.   (f) $F(0.99, 1.01)$.
(c) $F(1, 0)$.   (g) $F(1, 2)$.
(d) $F(0, 1)$.   (h) $F(4.9, 153.2)$.

(A diagram like Figure 7.3 may be helpful.)

8. An urn contains 8 white, 7 red, and 10 blue balls; 3 balls are drawn without replacement. Let X represent the number of white balls drawn and Y the number of red balls drawn.

(a) Find $p(1, 1)$, $p(2, 1)$, $p(0, 2)$, and $p(1, 0)$.
(b) Find $p(1)$, $q(2)$, where $p(x)$ and $q(y)$ are the marginal distributions for x and y, respectively.

(c) Find $q(y|x)$ for $x = 1$, $y = 2$; for $x = 2$, $y = 0$; for $x = 0$, $y = 3$; for $x = 3$, $y = 0$.

(d) Are X and Y independent?

9. Two dice, one red and one green, are tossed.

    (a) Construct a table showing the joint and marginal distributions for X and Y, the numbers showing on the red die and on the green die, respectively.

    (b) Are X and Y independent?

FORMULAS FOR JOINT PROBABILITIES. It is not always necessary to have the probabilities $p(x, y)$ *individually* specified. They may be given by a formula involving x and y, as in the following example.

    Example. A joint probability distribution for (X, Y) is given by

$$p(x, y) = (3x + 2y - 4)/54$$

for $x = 1$, 2, and 3 and $y = 1$, 2, and 3. The resulting joint probability table is shown in Table 7.3.

Table 7.3. Joint Probability Table Calculated from
$$p(x, y) = (3x + 2y - 4)/54, \; x, y = 1, 2, 3$$

	y			Marginals
x	1	2	3	$p(x)$
1	$\frac{1}{54}$	$\frac{1}{18}$	$\frac{5}{54}$	$\frac{1}{6}$
2	$\frac{2}{27}$	$\frac{1}{9}$	$\frac{4}{27}$	$\frac{1}{3}$
3	$\frac{7}{54}$	$\frac{1}{6}$	$\frac{11}{54}$	$\frac{1}{2}$
Marginals $q(y)$	$\frac{2}{9}$	$\frac{1}{3}$	$\frac{4}{9}$	1

FORMULAS FOR MARGINAL PROBABILITIES. The marginal probabilities can also be expressed as formulas if $p(x, y)$ is given as a formula. For example, in the example above, we find

$$p(x) = \sum_y p(x, y)$$

$$= \sum_{y=1}^{3} (3x + 2y - 4)/54$$

$$= 3(3x - 4)/54 + (2/54)\sum_{y=1}^{3} y$$

$$= \tfrac{1}{6}x - \tfrac{2}{9} + \tfrac{2}{9}$$

$$= \tfrac{1}{6}x.$$

Similarly we can show that

$$q(y) = \sum_{x=1}^{3} (3x + 2y - 4)/54$$
$$= \tfrac{1}{9}(y + 1).$$

Note 1. Since $p(x)q(y) = \tfrac{1}{54}x(y + 1)$ and this is *not* equal to $p(x, y)$, it is clear that X and Y are *not* independent in this example.

Note 2.

$$\sum_{x=1}^{3} p(x) = \sum_{x=1}^{3} \tfrac{1}{6}x = \tfrac{1}{6}(1 + 2 + 3) = 1,$$

$$\sum_{y=1}^{3} q(y) = \sum_{y=1}^{3} \tfrac{1}{9}(y + 1) = \tfrac{1}{9}(2 + 3 + 4) = 1,$$

as is also clear from Table 7.3.

FORMULAS FOR CONDITIONAL PROBABILITIES. The conditional probabilities can also be expressed as formulas if $p(x, y)$ is given as a formula. For example, in our most recent example, we have

$$q(y|x) = \frac{p(x, y)}{p(x)} = \frac{(3x + 2y - 4)/54}{x/6} = \frac{(3x + 2y - 4)}{9x},$$

$$p(x|y) = \frac{p(x, y)}{q(y)} = \frac{(3x + 2y - 4)/54}{(y + 1)/9} = \frac{(3x + 2y - 4)}{6(y + 1)}.$$

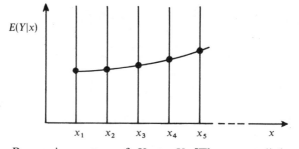

Figure 7.4. Regression curve of Y on X. [The curve links the values $E(Y|x_1)$, $E(Y|x_2)$, . . . , which are the conditional means of Y at $X = x_1$, $X = x_2$, ...]

Exercises

1. Draw a figure for the joint distribution given in Table 7.3.
2. If $p(x, y) = (2 + x + y)/33$, $x = 1, 2$, and $y = 1, 2, 3$, find the formulas for the marginal and conditional probabilities. Also draw a figure that shows the joint distribution.
3. If $p(x, y) = (x^2 - y)/20$, $x = 2, 3$, and $y = 1, 2$, find the formulas for the marginal and conditional probabilities. Also draw a figure that shows the joint distribution.

7.2. Regression Curves

Suppose two variables X and Y have a joint probability distribution function $p(x, y)$. Suppose we fix X at a value x. At this value $X = x$ the variable Y has a conditional distribution $q(y|x)$. This conditional distribution has a mean value defined by

$$E(Y|x) = \sum_y yq(y|x).$$

This is the *conditional mean* of Y at $X = x$ and depends, of course, on the value of x. We can now evaluate $E(Y|x)$ *for each possible value of* x and plot it, *as ordinate*, against x as abscissa. If all the points so plotted lie on a smooth curve we can call it, for convenience, *a regression curve* of Y on X. It is represented symbolically in Figure 7.4. Actually, for a *discrete* distribution, a regression curve of Y on X has real meaning only *at* the actual discrete values of x for which X is defined. (However, for continuous distributions, every point of the curve in the given range of x would be relevant.)

   Example 1. Consider the distribution of Table 7.1. First we find the three conditional means for $x = 1, 2$, and 3.
   (a) For $x = 1$,

$$q(y|1) = \frac{p(1, y)}{p(1)} = \begin{cases} \frac{1}{7}/\frac{2}{5} = \frac{5}{14} \text{ for } y = 1, \\ \frac{1}{7}/\frac{2}{5} = \frac{5}{14} \text{ for } y = 2, \\ \frac{4}{35}/\frac{2}{5} = \frac{2}{7} \text{ for } y = 3, \end{cases}$$

$$E(Y|1) = \sum yq(y|1) = 1(\tfrac{5}{14}) + 2(\tfrac{5}{14}) + 3(\tfrac{2}{7}) = \tfrac{27}{14}.$$

We plot this value at $x = 1$.

(b) For $x = 2$

$$q(y|2) = \frac{p(2, y)}{p(2)} = \begin{cases} \frac{1}{7}/\frac{1}{3} = \frac{3}{7} \text{ for } y = 1, \\ \frac{2}{21}/\frac{1}{3} = \frac{2}{7} \text{ for } y = 2, \\ \frac{2}{21}/\frac{1}{3} = \frac{2}{7} \text{ for } y = 3, \end{cases}$$

$$E(Y|2) = \sum yq(y|2) = 1(\tfrac{3}{7}) + 2(\tfrac{2}{7}) + 3(\tfrac{2}{7}) = \tfrac{13}{7}.$$

We plot this value at $x = 2$.

(c) For $x = 3$, $E(Y|3) = \frac{17}{7}$. (The reader should check the details of this calculation as an exercise.) We plot this value at $x = 3$. The plot of $E(Y|x)$ versus x is shown in Figure 7.5 and a regression curve is shown joining the points.

Example 2. When $p(x, y)$ is given as a function of x and y, points on a regression curve can also be obtained in terms of symbols as the following example shows. Suppose

$$p(x, y) = (3x + 2y - 4)/54 \qquad x, y = 1, 2, 3.$$

Then

$$\begin{aligned}
E(Y|x) &= \sum_{y=1}^{3} yq(y|x) \\
&= \sum_{y=1}^{3} \frac{y(3x + 2y - 4)}{9x} \\
&= \frac{1}{9x}\{(3x - 2) + 2(3x) + 3(3x + 2)\} \\
&= \frac{2(9x + 2)}{9x},
\end{aligned}$$

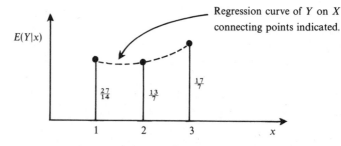

Regression curve of Y on X connecting points indicated.

$E(Y|x)$

$\frac{27}{14}$ $\frac{13}{7}$ $\frac{17}{7}$

1   2   3 x

Figure 7.5. Regression curve of Y on X for the joint distribution of Table 7.1.

which can be plotted against x. For the specific values of x involved we find that

$$E(Y|x) = \begin{cases} \frac{22}{9} & \text{for } x = 1, \\ \frac{20}{9} & \text{for } x = 2, \\ \frac{58}{27} & \text{for } x = 3. \end{cases}$$

A regression curve can be drawn through these three points if desired.

TWO REGRESSION CURVES. We have talked above about the regression of Y on X. There is also a regression of X on Y, in which, compared with the work above, the roles of X and Y and of x and y are interchanged. This interchange is not difficult once the work above has been properly grasped. We leave the details to the reader. (See Exercises 2 and 3 below.)

In general, the regression of Y on X is *quite different* from the regression of X on Y and (in general) two regression curves derived from the two calculations will differ.

STRAIGHT-LINE REGRESSION CURVES. It is possible for a regression curve to be a straight line. Although we do not provide any examples for the discrete case we remark that, for the bivariate normal distribution, which is an important continuous distribution, both the regression curves of Y on X and of X on Y are, in fact, straight lines.

Exercises

1. Check calculation (c) for Example 1 of this section.
2. Find the points $\{y, E(X|y)\}$ which lie on the regression curve of X on Y for Example 1. (See Table 7.1.)
3. For Example 2 of this section, where

$$p(x, y) = (3x + 2y - 4)/54 \qquad x, y = 1, 2, 3,$$

   show that

$$E(X|y) = (2y + 3)/(y + 1)$$

   and plot this against y for $y = 1, 2, 3$. Draw a regression curve through the three points.
4. Show that $E\{E(X|Y)\} = E(X)$. [*Note*: The inner expectation is taken over the conditional distribution of X with Y fixed at some specified value y. The outer expectation is taken over the *marginal* distribution of Y, $q(y)$. This is an important result.]
5. Show that, if X and Y are independent, $E(X|y) = EX$ and $E(Y|x) = EY$.

7.3. Bivariate Moments and Correlation

We have seen how the moments of a distribution of a single random variable X characterize the distribution. For example, the first moment about the origin, EX, is a measure of location, while the second moment about the mean, that is, the variance $V(X) = E(X - EX)^2 = EX^2 - (EX)^2$, is a measure of spread of the distribution.

BIVARIATE MOMENTS. We now define bivariate moments for a *joint* distribution of two random variables, X and Y. The general moment of order (p, q) about the point (a, b) is defined as

$$E(X - a)^p(Y - b)^q = \sum_x \sum_y (x - a)^p(y - b)^q p(xy)$$

for $p, q = 0, 1, 2, \ldots$, the summation being taken over all possible values of x and y.

MOMENTS ABOUT ZERO. If we set $a = 0$, $b = 0$, we obtain the moments about zero given by

$$\mu'_{pq} = EX^p Y^q = \sum_x \sum_y x^p y^q p(x, y).$$

In particular we note that

$$\mu'_{10} = EX = \sum_x \sum_y x p(x, y) = \sum_x x p(x),$$

$$\mu'_{01} = EY = \sum_x \sum_y y p(x, y) = \sum_y y q(y),$$

where $p(x)$, $q(y)$ are the appropriate marginal probabilities. Note that the notations EX and EY are quite unambiguous, because, for EX for example, we can merely take expectation with respect to the marginal $p(x)$ *or* with respect to $p(x, y)$, and the answers are identical.

MOMENTS ABOUT THE MEANS. If we set $a = \mu'_{10} = EX$, $b = \mu'_{01} = EY$, we obtain the moments about the means given by

$$\mu_{pq} = E(X - EX)^p(Y - EY)^q = \sum_x \sum_y (x - EX)^p(y - EY)^q p(x, y).$$

In particular, we note that

$$\mu_{10} = E(X - EX) = EX - EX = 0,$$

$$\mu_{01} = E(Y - EY) = EY - EY = 0,$$

$$\mu_{20} = E(X - EX)^2 = \sum_x \sum_y (x - EX)^2 p(x, y)$$

$$= \sum_x (x - EX)^2 p(x)$$

$$= V(X),$$

$$\mu_{02} = E(Y - EY)^2 = \sum_x \sum_y (y - EY)^2 p(x, y)$$

$$= \sum_y (y - EY)^2 q(y)$$

$$= V(Y),$$

$$\mu_{11} = E(X - EX)(Y - EY) = \sum_x \sum_y (x - EX)(y - EY) p(x, y)$$

$$= \mathrm{Cov}(X, Y).$$

The moment μ_{11} is called the *covariance* between X and Y, hence the notation $\mathrm{Cov}(X, Y)$. We recall that

$$V(X) = E(X - EX)^2 = EX^2 - (EX)^2.$$

By expanding the product we can write, in similar vein,

$$\mu_{11} = E(X - EX)(Y - EY)$$

$$= E\{XY - (EX)Y - (EY)X + (EX)(EY)\}$$

$$= E(XY) - (EX)(EY) - (EY)(EX) + (EX)(EY)$$

$$= E(XY) - (EX)(EY).$$

Now suppose that X and Y are independent random variables. Then by definition

$$p(x, y) = p(x)q(y)$$

and it follows that

$$E(XY) = \sum_x \sum_y xy p(x)q(y)$$

$$= \left\{ \sum_x x p(x) \right\} \left\{ \sum_y y q(y) \right\}$$

$$= (EX)(EY).$$

In this case, $\mu_{11} = 0$. So, *independent variables have zero covariance between them*. The converse result is not true, that is, the fact that $\mathrm{Cov}(X, Y) = 0$

does *not* necessarily imply that X and Y are independent, as the following example shows.

   Example.   Suppose the joint distribution of X and Y, $p(x, y)$, is defined by Table 7.4. Then clearly $EX = 0$, $E(XY) = 0$, so that $\text{Cov}(X, Y) = 0$. However, (for example) the product $p(2)q(3) = \frac{1}{64}$, while $p(2, 3) = 0$; so clearly it is *not* true that $p(x, y) = p(x)q(y)$; that is, the variables X and Y are *not* independent.

Table 7.4.   Joint Probability Table

		y		
x	1	2	3	$p(x)$
2	0	$\frac{1}{8}$	0	$\frac{1}{8}$
0	$\frac{1}{8}$	$\frac{1}{2}$	$\frac{1}{8}$	$\frac{3}{4}$
-2	0	$\frac{1}{8}$	0	$\frac{1}{8}$
$q(y)$	$\frac{1}{8}$	$\frac{3}{4}$	$\frac{1}{8}$	1

Exercises

1. For the distribution given by the following table, find
   (a) μ'_{10}.   (d) μ_{02}.
   (b) μ'_{01}.   (e) μ_{11}.
   (c) μ_{20}.

		y	
x	0	1	2
0	$\frac{1}{6}$	$\frac{1}{6}$	$\frac{1}{6}$
1	$\frac{1}{12}$	$\frac{1}{3}$	$\frac{1}{12}$

2. Repeat Exercise 1 for the distribution given in Table 7.1.
3. Repeat Exercise 1 for the distribution given in Table 7.2.
4. Repeat Exercise 1 for the distribution given in Table 7.3.
5. Repeat Exercise 1 for the distribution given in Table 7.4.
6. Repeat Exercise 1 for the distribution given by the following table:

			y	
x	1	2	3	4
1	$\frac{1}{9}$	$\frac{1}{6}$	$\frac{1}{18}$	$\frac{1}{9}$
2	$\frac{1}{18}$	$\frac{1}{9}$	$\frac{1}{18}$	$\frac{1}{3}$

7. Repeat Exercise 1 for the distribution $p(x, y) = (2 + x + y)/33$, $x = 1, 2$ and $y = 1, 2, 3$.

8. Repeat Exercise 1 for the distribution $p(x, y) = (x^2 - y)/20$, $x = 2, 3$ and $y = 1, 2$.

THE CORRELATION COEFFICIENT. The correlation coefficient, usually denoted by the Greek letter rho, written ρ, is a single, dimensionless measure of association between two random variables. The correlation coefficient between X and Y is defined as

$$\rho = \frac{\text{Cov}(X, Y)}{\{V(X)V(Y)\}^{1/2}} = \frac{\mu_{11}}{\{\mu_{20}\mu_{02}\}^{1/2}}.$$

It can be shown that $-1 \le \rho \le 1$ *always.* (Do Exercise 9 on page 165.) Note how ρ is dimensionless; if we multiply the values of x and/or y by any constant, the constant cancels out. (Do Exercise 10 on page 165.)

When ρ is close to 1, we say that X and Y have a high positive correlation, or are highly correlated in the positive direction. When ρ is close to -1, we say that X and Y have a high negative correlation, or are highly correlated in the negative direction.

When ρ is positive, the higher (or lower) values of X tend to be associated with the higher (or lower) values of Y in the following sense: The distribution $p(x, y)$ tends to have its higher values in regions where X and Y are *both* at their lower values or *both* at intermediate values or *both* at higher values. (Full account must be taken of sign in interpreting the words "lower" and "higher.")

Example. For Table 7.5, we can make the following calculations:

$$EX = 1(\tfrac{3}{16}) + 2(\tfrac{5}{8}) + 3(\tfrac{3}{16}) = 2,$$

$$EY = -7(\tfrac{3}{16}) + 0(\tfrac{5}{16}) + 4(\tfrac{5}{16}) + 6(\tfrac{3}{16}) = \tfrac{17}{16},$$

$$EX^2 = 1(\tfrac{3}{16}) + 4(\tfrac{5}{8}) + 9(\tfrac{3}{16}) = \tfrac{70}{16},$$

$$EY^2 = 49(\tfrac{3}{16}) + 0(\tfrac{5}{16}) + 16(\tfrac{5}{16}) + 36(\tfrac{3}{16}) = \tfrac{335}{16},$$

$$E(XY) = 1(-7)(\tfrac{1}{8}) + 1(0)(\tfrac{1}{16})$$

$$+ 2(-7)(\tfrac{1}{16}) + 2(0)(\tfrac{1}{4}) + 2(4)(\tfrac{1}{4}) + 2(6)(\tfrac{1}{16})$$

$$+ 3(4)(\tfrac{1}{16}) + 3(6)(\tfrac{1}{8})$$

$$= 4.$$

Table 7.5.   Joint Probability Table

x	y				$p(x)$
	-7	0	4	6	
1	$\frac{1}{8}$	$\frac{1}{16}$	0	0	$\frac{3}{16}$
2	$\frac{1}{16}$	$\frac{1}{4}$	$\frac{1}{4}$	$\frac{1}{16}$	$\frac{5}{8}$
3	0	0	$\frac{1}{16}$	$\frac{1}{8}$	$\frac{3}{16}$
$q(y)$	$\frac{3}{16}$	$\frac{5}{16}$	$\frac{5}{16}$	$\frac{3}{16}$	1

It follows that

$$\mu_{20} = EX^2 - (EX)^2 = \tfrac{70}{16} - 4 = \tfrac{3}{8},$$

$$\mu_{02} = EY^2 - (EY)^2 = \tfrac{335}{16} - \tfrac{289}{256} = \tfrac{5071}{256},$$

$$\mu_{11} = E(XY) - (EX)(EY) = 4 - \tfrac{17}{8} = \tfrac{15}{8},$$

$$\rho = \frac{\mu_{11}}{\{\mu_{20}\mu_{02}\}^{1/2}} = \frac{\frac{15}{8}}{\{\frac{3}{8} \times \frac{5071}{256}\}^{1/2}} = 0.688.$$

Note that the positive value of ρ reflects the fact that the higher probability values lie, roughly speaking, on a "ridge" passing from upper left to lower right of Table 7.5. If the ridge passed from lower left to upper right the correlation would be negative, as the following example indicates.

Table 7.6.   Joint Probability Table

x	y				$p(x)$
	-7	0	4	6	
1	0	0	$\frac{1}{16}$	$\frac{1}{8}$	$\frac{3}{16}$
2	$\frac{1}{16}$	$\frac{1}{4}$	$\frac{1}{4}$	$\frac{1}{16}$	$\frac{5}{8}$
3	$\frac{1}{8}$	$\frac{1}{16}$	0	0	$\frac{3}{16}$
$q(y)$	$\frac{3}{16}$	$\frac{5}{16}$	$\frac{5}{16}$	$\frac{3}{16}$	1

Example.   For the example of Table 7.6, the reader should confirm, for himself, the following results:

$$EX = 2 \qquad EY = \tfrac{17}{16},$$

$$EX^2 = \tfrac{70}{16} \qquad EY^2 = \tfrac{335}{16} \qquad E(XY) = 0,$$

$$\mu_{20} = \tfrac{3}{8} \qquad \mu_{02} = \tfrac{5071}{256} \qquad \mu_{11} = -\tfrac{17}{8},$$

$$\rho = -0.780.$$

ADDITIONAL COMMENTS. The correlation coefficient ρ is a measure of *linear* association.

If there is a perfect linear relationship between X and Y, that is, if $Y = aX + b$, where a and b are constants, then $\rho = 1$, or $\rho = -1$, depending upon whether a is positive or negative. (Do Exercise 8, page 165.) The joint distribution would, in this case, consist only of points on the line $Y = aX + b$.

A value of ρ close to zero implies that there is little tendency for X and Y to be associated in a linear manner.

Note. *Independent variables X and Y have zero correlation $\rho = 0$. This is because, if X and Y are independent, $\mu_{11} = 0$, as shown above, and hence $\rho = 0$.*

The converse result is *not* true; that is, the fact that $\rho = 0$ does *not* necessarily imply that X and Y are independent. [For a counterexample, re-examine the example in which $\text{Cov}(X, Y) = 0$ in Table 7.4.] The reason for this is that, as mentioned above, ρ is a measure of *linear* association, and two variables not associated in a linear manner may be related in some other manner.

Note. For the bivariate normal distribution, which is a continuous distribution and beyond the scope of this book, $\rho = 0$ *does* imply independence. We note this so that the reader can appreciate that this *can* occur although, more generally, it does not occur for the majority of distributions.

We now provide an example of finding the correlation coefficient ρ when $p(x, y)$ is given by a formula involving both x and y.

Example. Consider again the joint distribution for X and Y given by the joint probability distribution function

$$p(x, y) = (3x + 2y - 4)/54 \qquad x = 1, 2, 3; y = 1, 2, 3,$$

and let us find the value of the correlation coefficient ρ. We must first find the variances of the individual variables, that is, of the marginal distributions, and also the covariance. We shall find, in succession, μ'_{10}, μ'_{01}, μ_{20}, μ_{02}, and μ_{11}, as follows:

$$EX = \mu'_{10} = \sum\sum x(3x + 2y - 4)/54$$
$$= \sum_x [x(3x + 2 - 4) + x(3x + 4 - 4) + x(3x + 6 - 4)]/54$$
$$= \sum_x x(9x)/54 = \tfrac{7}{3},$$

$$EY = \mu'_{01} = \sum\sum y(3x + 2y - 4)/54$$
$$= \sum_y [y(3 + 2y - 4) + y(6 + 2y - 4) + y(9 + 2y - 4)]/54$$
$$= \sum_y y(6y + 6)/54 = \tfrac{20}{9},$$

$$\mu_{20} = E(X - \tfrac{7}{3})^2 = \sum\sum (x - \tfrac{7}{3})^2(3x + 2y - 4)/54$$
$$= \sum\sum [x^2(3x + 2y - 4) - (14x/3)(3x + 2y - 4)$$
$$+ (\tfrac{49}{9})(3x + 2y - 4)]/54$$
$$= \sum_x [9x^3 - (14x/3)9x^2 + (\tfrac{49}{9})9x]/54 = \tfrac{5}{9},$$

$$\mu_{02} = E(Y - \tfrac{20}{9})^2 = \sum\sum (y^2 - 40y/9 + \tfrac{400}{81})(3x + 2y - 4)/54$$
$$= \sum_y [y^2(6y + 6) - (40y/9)(6y + 6) + (\tfrac{400}{81})(6y + 6)]/54$$
$$= \tfrac{50}{81},$$

$$\mu_{11} = E(X - \tfrac{7}{3})(Y - \tfrac{20}{9}) = \sum\sum (x - \tfrac{7}{3})(y - \tfrac{20}{9})(3x + 2y - 4)/54$$
$$= \sum\sum (xy - 7y/3 - 20x/9 + \tfrac{140}{27})(3x + 2y - 4)/54$$
$$= \left\{\sum_x (\tfrac{140}{27} - 20x/9)(9x) - \sum_y 7y(6y + 6)/3\right.$$
$$\left. + \sum\sum xy(3x + 2y - 4)\right\} \Big/ 54$$
$$= -\tfrac{2}{27}.$$

Hence

$$\rho = (-\tfrac{2}{27})/\sqrt{(\tfrac{5}{9})(\tfrac{50}{81})} = -\sqrt{\tfrac{10}{25}} = -0.126.$$

This result confirms the conclusion reached, in Section 7.1, page 154, that X and Y are not independent, for, if they were independent, ρ would have to be zero. (Note carefully, however, that if ρ had been zero we could *not* have concluded that X and Y *were* independent, as mentioned above.)

Exercises

1. Find ρ for the following bivariate distribution:

		y	
x	0	1	2
0	$\tfrac{1}{6}$	$\tfrac{1}{6}$	$\tfrac{1}{6}$
1	$\tfrac{1}{12}$	$\tfrac{1}{3}$	$\tfrac{1}{12}$

2. Find ρ for the distribution of Table 7.1.
3. Find ρ for the distribution of Table 7.2.
4. Find ρ for the distribution of Table 7.3.

5. Find ρ for the distribution given by the following table.

x	y			
	1	2	3	4
1	$\frac{1}{9}$	$\frac{1}{6}$	$\frac{1}{18}$	$\frac{1}{9}$
2	$\frac{1}{18}$	$\frac{1}{9}$	$\frac{1}{18}$	$\frac{1}{3}$

6. Find ρ for the distribution $p(x, y) = (2 + x + y)/33$, $x = 1, 2$ and $y = 1, 2, 3$.
7. Find ρ for the distribution $p(x, y) = (x^2 - y)/20$, $x = 2, 3$ and $y = 1, 2$.
8. Show that, if $Y = aX + b$, then $\rho = 1$ for $a > 0$, and $\rho = -1$ for $a < 0$. What is ρ if $a = 0$?
9. Show that $-1 \le \rho \le 1$. [*Hints*: (a) Consider the inequalities

$$E\left\{\frac{X - EX}{\{V(X)\}^{1/2}} \pm \frac{Y - EY}{\{V(Y)\}^{1/2}}\right\}^2 \ge 0$$

and expand the left-hand side, first for the plus sign, then for the minus sign; this assumes $V(X) \ne 0$, $V(Y) \ne 0$. If $V(X) = 0$ and/or $V(Y) = 0$, $(X - EX)(Y - EX) = 0$ with probability 1 (why?) and so $\rho = 0$, and the result is clearly true in this case. (b) An alternative method is to consider the inequality

$$f(z) = E\{z(X - EX) + (Y - EY)\}^2 \ge 0,$$

which is true for any real z, and to write down the condition $B^2 \le AC$ that $f(z) = Az^2 + 2Bz + C = 0$ has equal or imaginary roots.]
10. Suppose $X' = \alpha X$ and $Y' = \beta Y$. Show that the correlation between X' and Y' is identical to that between X and Y, confirming that ρ is a dimensionless quantity.

7.4. Multinomial Distribution

One of the three conditions for the valid application of the binomial probability law is that all possible outcomes of a probabilistic phenomenon can be classified either as successes *or* failures. (Can you remember the other two conditions?) When there are *more* than two categories of classification, we need the *multinomial distribution.*

Let us consider the simple experiment of tossing a single die, and focus attention on the number of dots showing on the top face. Clearly, six mutually exclusive outcomes are possible. Let the probabilities assigned to these outcomes be $p_1, p_2, \ldots, p_6$ (for a fair die all these would be equal to $\frac{1}{6}$), and let us make n independent trials (that is, make n tosses). The probability that the overall result will be x_1 ones, x_2 twos, $\ldots, x_6$ sixes *in a particular order* is given by

$$(p_1^{x_1})(p_2^{x_2}) \cdots (p_6^{x_6}).$$

The number of possible arrangements of x_1 ones, x_2 twos, $\ldots, x_6$ sixes is just the number of permutations of n objects when there are x_1 of one kind, x_2 of a second kind, $\ldots, x_6$ of a sixth kind, which is

$$\frac{n!}{x_1!\,x_2!\,x_3!\,x_4!\,x_5!\,x_6!}.$$

(This is true because there are, altogether, $n!$ ways of arranging n objects, but the $x_i!$ ways of arranging the x_i objects of the ith kind are identical for our purposes.) It follows that the probability of obtaining x_1 ones, x_2 twos, $\ldots,$ x_6 sixes is

$$p(x_1, x_2, x_3, x_4, x_5, x_6) = \frac{n!}{x_1!\,x_2!\cdots x_6!}p_1^{x_1}p_2^{x_2} \cdots p_6^{x_6}.$$

Special case. If $p_1 = p_2 = \cdots = p_6 = \frac{1}{6}$, the probability of obtaining one of each number in six tosses of a fair die is given by

$$\frac{6!}{1!\,1!\cdots 1!}\left(\frac{1}{6}\right)^1\left(\frac{1}{6}\right)^1 \cdots \left(\frac{1}{6}\right)^1 = \frac{6!}{6^6} = \frac{5}{324}.$$

GENERAL FORMULA FOR MULTINOMIAL PROBABILITIES. Suppose that a probabilistic phenomenon has k possible outcomes with probabilities $p_1, p_2, \ldots, p_k$ (where $\sum p_i = 1$). Suppose also that these probabilities are constant for every trial and that all trials are independent. Then the probability of x_1 outcomes of the first kind, x_2 outcomes of the second kind, $\ldots, x_k$ outcomes of the kth kind, in n trials, where

$$x_1 + x_2 + \cdots + x_k = n,$$

is given by

$$p(x_1, x_2, \ldots, x_k) = \frac{n!}{x_1!\,x_2!\cdots x_k!}p_1^{x_1}p_2^{x_2} \cdots p_k^{x_k}.$$

This expression is called a *multinomial probability*. We recall that the binomial probability law represents the general term in the expansion of a binomial expression. Similarly, the multinomial law represents the general term in the expansion of the multinomial expression $(p_1 + p_2 + \cdots + p_k)^n$. Note that, since $\sum x_i = n$, it is *impossible* for the variables $X_1, X_2, \ldots, X_k$ (whose attained values are $x_1, x_2, \ldots, x_k$) to be independent, ever. The distribution is rather more complex than anything we have examined to this point.

SOME DERIVED RESULTS. We state the following results without proof and ask the reader to accept them.

1. The marginal distribution of X_i is binomial with parameters n and p_i, for all $i = 1, 2, \ldots, k$.

2. Given that $X_j = v_j$ (fixed), the (conditional) marginal distribution of X_i is binomial with parameters "n" $= n - v_j$ and "p" $= p_i/(1 - p_j)$, provided that $p_j < 1$, for all $i, j = 1, 2, \ldots, k, i \neq j$.

3. $\text{Cov}(X_i X_j) = -n p_i p_j$, for all $i, j = 1, 2, \ldots, k, i \neq j$.

4. The correlation between X_i and X_j is given by

$$\rho_{ij} = -\left\{\frac{p_i p_j}{(1 - p_i)(1 - p_j)}\right\}^{1/2}.$$

Exercises

1. Find the probability of getting 3 fours, 2 twos, and 1 six in (a) 6 and (b) 10 tosses of a die.

2. A box contains 20 red, 30 white, and 50 blue balls. If 10 balls are withdrawn, one at a time, *with replacement*, find the probability of getting x red, y white, and z blue balls.

3. Five cards are drawn one at a time, *with replacement*, from a bridge deck. Find the probability of getting 2 spades, 1 heart, 1 diamond, and 1 club.

4. Repeat Exercise 8, page 152, except that, this time, consider the balls to be drawn *with replacement*.

5. A box contains 10 balls; 3 are black, 3 are red, and 4 are white. Four balls are drawn successively *with replacement*, and X_1, X_2, and X_3 denote respectively, the numbers of black, red, and white balls among the four drawn.

   (a) Evaluate the probabilities for all possible values x_1, x_2, and x_3, the numbers of black, red, and white balls among the four drawn.

   (b) Find the marginal distributions of X_1, X_2, and X_3.

   (c) Find the (conditional) marginal distributions of (1) X_1 given that $X_3 = 0$, (2) X_2 given $X_1 = 2$, (3) X_3 given $X_2 = 1$.

   (d) Find $\rho_{12}, \rho_{13}, \rho_{23}$ where ρ_{ij} denotes the correlation between X_i and X_j.

6. Show that, for a multinomial distribution of X_1, X_2, X_3, and X_4 with parameters n, p_1, p_2, p_3, and p_4, the conditional distribution of X_1, X_2, and X_3 given that $X_4 = v_4$ is multinomial with parameters $n - v_4$, $p_1/(1 - p_4)$, $p_2/(1 - p_4)$, and $p_3/(1 - p_4)$. Write down what you think is the generalization of this result for $X_1, X_2, \ldots, X_k$ when $X_{k-1} = v_{k-1}$ and $X_k = v_k$.

7. Of the n people present at church gatherings in a particular church, 10% are usually children and 50% are usually male. If, at one gathering, $n = 10$, what is the probability of there being
   (a) three boys, two girls, and five adults?
   (b) five men, two women, and three children?

8. Seven secretaries are having coffee together and are comparing their birth dates. What is the probability that two have birthdays in January, February, or March; two have birthdays in December; and three have birthdays in the remaining months. (Assume that a birthday is equally likely to occur in any month, as an approximation to the truth.)

9. Four students in Madison, Wisconsin, are copying each other's homework. What is the probability that more of these students have their birthdays in (1) March, April, May, or July than in (2) June, August, September, or October. [*Hint*: Divide the year into three periods consisting of I (Mar., Apr., May, July), II (June, Aug., Sept., Oct.), III (Jan., Feb., Nov., Dec.) each with probability $\frac{1}{3}$. We are interested in allocations of the four students to the three periods such that "students in I" exceed "students in II." Add the probabilities of the allocations (4, 0, 0), (3, 1, 0), (3, 0, 1), (2, 1, 1), (2, 0, 2), (1, 0, 3) to get $\frac{31}{81}$.]

10. If $X_1, X_2, \ldots, X_n$ are multinomial with parameters $n, p_1, p_2, \ldots, p_n$, what are EX_i and $V(X_i)$ for $i = 1, 2, \ldots, n$? (*Hint*: Use the first derived result on page 167.)

11. Show that, in the derived results on page 167, results 1 and 3 imply result 4.

12. Use derived result 2 on page 167 to show that for all $i, j = 1, 2, \ldots, k$, $i \neq j$,

$$E(X_i | X_j = v_j) = \frac{(n - v_j)p_i}{1 - p_j}.$$

13. A box has 5 red, 4 blue, 2 white, and 7 yellow marbles. Eight marbles are drawn at random from the box.
   (a) What is the probability of drawing 2 red, 2 blue, 1 white, and 3 yellow marbles if each marble is replaced before another is drawn?
   (b) What is the probability of drawing 2 red, 2 blue, 1 white, and 3 yellow marbles if the marbles are *not* replaced?

CHAPTER 8

Sums and Averages of Random Variables

8.1. Probability Functions for the Sum of Two or More Independent Random Variables

Suppose a pair of independent random variables X and Y have a joint probability distribution given by $p(x, y)$ as in Table 8.1. Note that, since the variables *are* independent,

$$p(x, y) = p(x)q(y).$$

Table 8.1. Joint Probability Function for Independent Variables X and Y

x	y				$p(x)$
	0	1	2	3	
0	$\frac{1}{56}$	$\frac{1}{14}$	$\frac{1}{28}$	$\frac{1}{56}$	$\frac{1}{7}$
1	$\frac{1}{28}$	$\frac{1}{7}$	$\frac{1}{14}$	$\frac{1}{28}$	$\frac{2}{7}$
2	$\frac{1}{14}$	$\frac{2}{7}$	$\frac{1}{7}$	$\frac{1}{14}$	$\frac{4}{7}$
$q(y)$	$\frac{1}{8}$	$\frac{1}{2}$	$\frac{1}{4}$	$\frac{1}{8}$	1

Suppose we wish to know the probability that $Z = X + Y = 3$. When the variables are independent we get this probability by considering all the values of x and y such that $x + y = 3$ and adding the corresponding probabilities $p(x, y)$ as follows:

x	y	$p(x, y)$
0	3	$\frac{1}{56}$
1	2	$\frac{1}{14}$
2	1	$\frac{2}{7}$

$$P(X + Y = 3) = \tfrac{3}{8}$$

Note that in doing this calculation we add together the probabilities on a certain diagonal of Table 8.1. Now clearly $Z \equiv X + Y$ can take all values from 0 to 5 and we can go through a number of "diagonal sum" calculations similar to the one above to get a table as follows, where $p(z) = P(Z = z)$:

z	0	1	2	3	4	5
$p(z)$	$\frac{1}{56}$	$\frac{3}{28}$	$\frac{1}{4}$	$\frac{3}{8}$	$\frac{5}{28}$	$\frac{1}{14}$

The reader should verify this table now; he should also verify that, as expected,

$$\sum_{z=0}^{5} p(z) = 1.$$

THE GENERAL RESULT.   In general, if X and Y are independent, and since the events in the union are independent, we have that

$$p(z) = P(X + Y = z) = \sum_{x} \sum_{y} P(X = x)P(Y = y),$$

where the summation is over all possible values of x and y such that $x + y = z$. If we write $y = z - x$, we get

$$p(z) = \sum_{x} P(X = x)P(Y = z - x)$$
$$= \sum_{x} p(x)q(z - x),$$

the summation being over all possible values of x.

[It is understood, of course, that $q(z - x) = 0$ if $z - x$ is not a value for which $q(z - x)$ is defined. In the example above, when $z = 5$ and $x = 1$,

$q(z - x) = q(4)$, which is not defined and so $q(4) = 0$. Also $q(5) = 0$. Thus

$$p(5) = p(0)q(5) + p(1)q(4) + p(2)q(3)$$
$$= p(2)q(3)$$
$$= \tfrac{1}{14}.]$$

An alternative definition is

$$p(z) = \sum_y p(z - y)q(y),$$

the summation being over all possible values of y and where $p(z - y) = 0$ if $z - y$ is not a value for which $p(z - y)$ is defined. This alternative definition arises from putting $x = z - y$ in the double summation for $p(z)$ given above.

EXTENSION TO MORE THAN TWO VARIABLES. The general result above can be applied repeatedly to find the probability distribution of the sum of any number of independent random variables X_i. For example, to find the probability distribution of $X_1 + X_2 + X_3 + X_4$, we find, successively, the distributions of

$$X_1 + X_2,$$
$$(X_1 + X_2) + X_3,$$

and

$$\{(X_1 + X_2) + X_3\} + X_4.$$

[A number of random variables $X_1, X_2, \ldots, X_n$ are independent if and only if their joint probability function can be expressed as a product of their separate marginal probability functions, that is, if and only if

$$p(x_1, x_2, \ldots, x_n) = p(x_1)q(x_2)\cdots r(x_n),$$

where $p, q, \ldots, r$ denote the various marginal probability functions.]

Example. A fair die is thrown n times in succession, and the random variable X_i is the number appearing on the ith toss. Let $S_n = X_1 + X_2 + \cdots + X_n$; the distribution of S_1 or of *any* X_i is as follows:

s_1	1	2	3	4	5	6	x_i
$p(s_1)$	$\tfrac{1}{6}$	$\tfrac{1}{6}$	$\tfrac{1}{6}$	$\tfrac{1}{6}$	$\tfrac{1}{6}$	$\tfrac{1}{6}$	$p(x_i)$

The joint distribution of $S_1 (= X_1)$ and X_2 is a 6 by 6 table with borders $1, 2, \ldots, 6$ and with all entries $\tfrac{1}{36}$. Obviously, X_1 and X_2 are independent.

Summing along the appropriate diagonals we get, for example,

$$
\begin{aligned}
P(S_2 \equiv X_1 + X_2 = 6) &= P(X_1 = 1)P(X_2 = 5) + P(X_1 = 2)P(X_2 = 4) \\
&\quad + P(X_1 = 3)P(X_2 = 3) + P(X_1 = 4)P(X_2 = 2) \\
&\quad + P(X_1 = 5)P(X_2 = 1) \\
&= \tfrac{5}{36}.
\end{aligned}
$$

The whole distribution of S_2 is as follows.

S_2	2	3	4	5	6	7	8	9	10	11	12
$p(s_2)$	$\frac{1}{36}$	$\frac{2}{36}$	$\frac{3}{36}$	$\frac{4}{36}$	$\frac{5}{36}$	$\frac{6}{36}$	$\frac{5}{36}$	$\frac{4}{36}$	$\frac{3}{36}$	$\frac{2}{36}$	$\frac{1}{36}$

From the table of the distribution of S_1 (or X_i) above and the table of the distribution of S_2 it is easy to get the distribution of $S_3 = S_2 + X_3$. For example, S_3 can take the value 12 via any of the following possibilities:

Value of S_2	11	10	9	8	7	6
Value of X_3	1	2	3	4	5	6

Taking the product (and then the overall sum) of appropriate probabilities, we get

$$
\begin{aligned}
P(S_3 = 12) &= \tfrac{1}{36} \cdot \tfrac{1}{6}\{2 \cdot 1 + 3 \cdot 1 + 4 \cdot 1 + 5 \cdot 1 + 6 \cdot 1 + 5 \cdot 1\} \\
&= \tfrac{25}{216}.
\end{aligned}
$$

The whole probability table is as follows:

S_3	3	4	5	6	7	8	9	10
$p(s_3)$	$\frac{1}{216}$	$\frac{3}{216}$	$\frac{6}{216}$	$\frac{10}{216}$	$\frac{15}{216}$	$\frac{21}{216}$	$\frac{25}{216}$	$\frac{27}{216}$
S_3	18	17	16	15	14	13	12	11

Note that, in this table $p(s_3)$ is below s_3 if s_3 is in the top line, and above s_3 if s_3 is in the bottom line. For example,

$$
p(6) = p(15) = \tfrac{10}{216},
$$

and so on.

Exercises

1. Suppose X_1, X_2, and X_3 are independent random variables each with the same probability function which assigns probabilities as follows:

$$
P(X_i = 0) = \tfrac{1}{3}, \quad P(X_i = 1) = \tfrac{1}{3}, \quad P(X_i = 2) = \tfrac{1}{3}.
$$

Find the probability distributions of
- (a) $X_1 + X_2$
- (b) $X_2 + X_3$
- (c) $X_1 + X_3$
- (d) $X_1 + X_2 + X_3$

and check that the probabilities add to one in every case.
2. Repeat Exercise 1 but with

$$P(X_i = 0) = \tfrac{1}{6}, \quad P(X_i = 1) = \tfrac{1}{3}, \quad P(X_i = 2) = \tfrac{1}{2}.$$

3. X and Y are independent random variables with probability distributions as follows:

x	-2	-1	0	1	2
$p(x)$	$\frac{1}{16}$	$\frac{1}{8}$	$\frac{3}{16}$	$\frac{1}{4}$	$\frac{3}{8}$

y	-5	-3	-1	1	3	5
$q(y)$	$\frac{1}{20}$	$\frac{1}{10}$	$\frac{1}{5}$	$\frac{2}{5}$	$\frac{3}{20}$	$\frac{1}{10}$

Find the probability distribution of $X + Y$, and check that your probabilities add to one.
4. Consider the die example in this section, and suppose we have the probability distributions of S_2 and S_3. How can we use these to get the probability distribution of S_5? Explain *why* in your answer.

8.2. Means and Variances of Sums and Averages of Random Variables

We now give some useful results for sums and averages of a number of random variables. Note that, in this section, the variables do not have to be independent, nor do we have to know the actual distributions of the sums and averages involved.

RESULTS FOR SUMS OF RANDOM VARIABLES. Suppose that $X_1, X_2, \ldots, X_n$ are n discrete random variables. Let $S_n = X_1 + X_2 + \cdots + X_n$. Then
1. $ES_n = EX_1 + EX_2 + \cdots + EX_n.$
2. $V(S_n) = V(X_1) + V(X_2) + \cdots + V(X_n) + 2 \operatorname{Cov}(X_1, X_2)$
$\qquad + 2 \operatorname{Cov}(X_1, X_3) + \cdots + 2 \operatorname{Cov}(X_{n-1}, X_n).$

The proofs of these statements are not difficult and we deal with the case $n = 2$ below. For larger n, the proofs are similar but simply need more developing and writing out.

Proofs for $n = 2$

1. $ES_n = E(X_1 + X_2)$

$$= \sum_{x_1} \sum_{x_2} (x_1 + x_2)p(x_1, x_2)$$

$$= \sum_{x_1} \sum_{x_2} x_1 p(x_1, x_2) + \sum_{x_1} \sum_{x_2} x_2 p(x_1, x_2)$$

$$= EX_1 + EX_2.$$

2. $V(S_n) = E\{S_n - ES_n\}^2$

$$= E\{X_1 + X_2 - EX_1 - EX_2\}^2$$

$$= E\{(X_1 - EX_1) + (X_2 - EX_2)\}^2$$

$$= E\{(X_1 - EX_1)^2 + (X_2 - EX_2)^2 + 2(X_1 - EX_1)(X_2 - EX_2)\}$$

$$= E(X_1 - EX_1)^2 + E(X_2 - EX_2)^2$$

$$+ 2E\{(X_1 - EX_1)(X_2 - EX_2)\}$$

$$= V(X_1) + V(X_2) + 2\,\mathrm{Cov}(X_1, X_2).$$

Special case. When $\mathrm{Cov}(X_i, X_j) = 0$ for all $i, j = 1, 2, \ldots, n$, result 1 above is still true and result 2 reduces to

3. $V(S_n) = V(X_1) + V(X_2) + \cdots + V(X_n).$

Note that if all the X's are independent, $\mathrm{Cov}(X_i, X_j) = 0$ and result 3 applies. (Note, however, that *if* result 3 is true, the X's are not necessarily independent. They are merely *uncorrelated* pairwise.)

RESULTS FOR AVERAGES OF RANDOM VARIABLES.   Let us write

$$\overline{X}_n = \frac{1}{n}S_n = \frac{1}{n}(X_1 + X_2 + \cdots + X_n)$$

for the average of n variables. Then

4. $E\overline{X}_n = \dfrac{1}{n}(EX_1 + EX_2 + \cdots + EX_n).$

5. $V(\overline{X}_n) = \dfrac{1}{n^2} V(S_n).$

These results follow easily from proofs similar to those of results 1 and 2 above. (Also see page 119, Example 3.)

Special cases.   Suppose all the X's have the same mean μ_x and the same variance σ_x^2, and the covariance between any two X's is $\rho\sigma_x^2$. That is, suppose that

$$EX_i = \mu_x,$$

$$V(X_i) = \sigma_x^2,$$

$$\text{Cov}(X_i, X_j) = \rho\sigma_x^2$$

are the same for all $i, j = 1, 2, \ldots, n$. Results 4 and 5 would then reduce to

6. $EX_n = \dfrac{1}{n}(\mu_x + \mu_x + \cdots + \mu_x)$

    $= \mu_x.$

7. $V(\overline{X}_n) = \dfrac{1}{n^2}\{\sigma_x^2 + \cdots + \sigma_x^2 + 2\rho\sigma_x^2 + \cdots + 2\rho\sigma_x^2\}$

    $= \dfrac{\sigma_x^2}{n} + \dfrac{n-1}{n}\rho\sigma_x^2$

    $= \dfrac{\sigma_x^2}{n}\{1 + (n-1)\rho\}.$

[This is because there are n of the σ_x^2 terms and $\dbinom{n}{2} = \frac{1}{2}n(n-1)$ of the $2\rho\sigma_x^2$ terms, the number of ways of selecting two values i and j from the possibilities $i, j = 1, 2, \ldots, n, i \neq j$.] Moreover, when $\rho = 0$, we have

8. $V(\overline{X}_n) = \dfrac{\sigma_x^2}{n}.$

(*Note:* If all X's have the same distribution and are independent, or simply uncorrelated pairwise, results 6 and 8 apply.)

Example.   A fair die is thrown 10 times and the average result $\overline{X}_{10}$ is recorded. What is the mean and variance of this random variable?

First we note that, if X_i is the result of the ith throw, X_i has the distribution which allots probability $1/6$ to each number 1, 2, 3, 4, 5, and 6. For each of these variables X_i we have

$$\mu_x = EX_i = \tfrac{1}{6}(1 + 2 + 3 + 4 + 5 + 6) = \tfrac{7}{2},$$

$$EX_i^2 = \tfrac{1}{6}(1 + 4 + 9 + 16 + 25 + 36) = \tfrac{91}{6},$$

$$V(X_i) = \tfrac{91}{6} - \tfrac{49}{4} = \tfrac{35}{12}.$$

Since the X_i are clearly independent, we can apply results 6 and 8 to give

$$EX_{10} = \tfrac{7}{2},$$
$$V(\overline{X}_{10}) = \tfrac{1}{10}(\tfrac{35}{12}) = \tfrac{7}{24}.$$

Exercises

1. A fair coin is tossed k times and S_k is the number of heads that occur. Find ES_k and $V(S_k)$. (*Hint:* Let X_i, $i = 1, 2, \ldots, k$ be binomial with $n = 1$, $p = \tfrac{1}{2}$, and values $x_i = 1, 0$ for heads and tails, respectively; that is, X_i is the result of a Bernoulli trial. Note that $S_k = X_1 + X_2 + \cdots + X_k$ and apply the results above. How else could this be done?)

2. The variables $X_1, X_2, \ldots, X_n$ are independent and identically distributed with probability distribution as follows:

x	-1	0	1	2
$p(x)$	$\tfrac{1}{7}$	$\tfrac{2}{7}$	$\tfrac{3}{7}$	$\tfrac{1}{7}$

If $\overline{X} = (X_1 + X_2 + \cdots + X_n)/n$, find (in terms of n), $E\overline{X}$ and $V(\overline{X})$.

3. If $X_1, X_2, \ldots, X_n$ are Poisson variables with parameters $\lambda_1, \lambda_2, \ldots, \lambda_n$, respectively, then (it is a fact which we do not prove in this book) $S_n = X_1 + X_2 + \cdots + X_n$ is a Poisson variable with parameter λ, say.
   (a) What is λ in terms of $\lambda_1, \lambda_2, \ldots, \lambda_n$?
   (b) What is the distribution of $\overline{X}_n = S_n/n$?

4. Let X_1 and X_2 be independent binomial random variables with parameters (n_1, p_1) and (n_2, p_2), respectively.
   (a) If $Y = X_1 + X_2$, find EY and $V(Y)$.
   (b) Suppose $n_1 = n_2 = 3$, $p_1 = \tfrac{1}{4}$, and $p_2 = \tfrac{1}{2}$. Write down the probability distributions of X_1 and X_2 and hence get the distribution of $Y = X_1 + X_2$. Is the distribution of Y binomial? What would be your conclusion from this for the distribution of the sum of k independent binomial variables $X_1, X_2, \ldots, X_k$?
   (c) Repeat (b) but with $n_2 = 2$ and everything else the same. What is your conclusion now?

5. Suppose $X_1, X_2, \ldots, X_k$ are multinomially distributed with parameters n and $p_1, p_2, \ldots, p_k$. The distribution of the sum of any subset of the k variables is binomial (a fact that we state without proof). What are the parameters of the distribution of $X_1 + X_2 + X_3$? (*Hint:* Use results 1 and 2 above and solve the two equations that result for the parameters of the binomial distribution.)

6. Suppose $X_1, X_2, \ldots, X_k$ are independent variables each with the geometric distribution with parameter p, that is, each with

$$p(x) = (1 - p)^{x-1}p \qquad x = 1, 2, \ldots.$$

(a) What is the distribution followed by $S_k = X_1 + X_2 + \cdots + X_k$? [*Hints:* (1) Find the distributions of S_2 and S_3 and see if you can mentally generalize. (2) Think of X_1 as being the number of binomial trials up to a first success, X_2 the number up to a second success, and so on ..., so that S_k is the *total* number of trials up to the kth success. What is the distribution of S_k now?]

(b) What are EX and $V(X)$ for the geometric distribution? Hence use results 1 and 3 of this section to find ES_k and $V(S_k)$.

7. Let $A = a_1 X_1 + a_2 X_2$, $B = b_1 X_1 + b_2 X_2$, where the X's denote random variables and the a's and b's are specified constants. Show that the covariance of the two random variables A and B is given by

$$\text{Cov}(A, B) = a_1 b_1 V(X_1) + a_2 b_2 V(X_2) + (a_1 b_2 + a_2 b_1) \text{Cov}(X_1, X_2).$$

8. Let $A = a_1 X_1 + a_2 X_2 + \cdots + a_n X_n$, $B = b_1 X_1 + b_2 X_2 + \cdots + b_n X_n$, where the X's denote random variables and the a's and b's are specified constants. Show that the covariance of the two random variables A and B is given by

$$\text{Cov}(A, B) = \sum_{i=1}^{n} a_i b_i V(X_i) + \sum_{\substack{i=1 \\ i \neq j}}^{n} \sum_{j=1}^{n} a_i b_j \text{Cov}(X_i, X_j).$$

9. By putting $a_i = b_i$ in the results of Exercises 7 and 8, find $V(A)$ in the two cases.

10. A fair six-sided die is tossed n times. Let X equal the number of ones and sixes observed, and let Y equal the number of fives and sixes observed. Show that the correlation ρ between X and Y *does not depend* on n and, in fact, that $\rho = \frac{1}{4}$. [*Hint:* If X_1 is the number of ones, ..., X_6 is the number of sixes, then the X_i's are multinomial with all $p_i = \frac{1}{6}$. Let $X = X_1 + X_6$, $Y = X_5 + X_6$. Get $V(X)$, $V(Y)$ from result 2 of this section (or from Exercise 9) and $\text{Cov}(X, Y)$ from Exercise 7, setting all $a_i = b_j = 1$ and using the multinomial results given in Section 7.4.]

11. A fair six-sided die is tossed n times. Let X equal the number of ones, twos, and sixes and Y equal the number of fours, fives, and sixes observed. Find the correlation between X and Y, and show that it does not depend on n. (*Hint:* Use the obvious extension of the method indicated for Exercise 10.)

12. Each of the independent random variables $X_1, X_2, \ldots, X_n$ follows the Poisson distribution

$$p(x) = e^{-\lambda} \lambda^x / x! \qquad x = 0, 1, 2, \ldots.$$

If $Y = X_1 + X_2 + \cdots + X_n$, find $P(Y = 0)$, $P(Y = 1)$, and $P(Y = 2)$, and confirm that they obey the general formula

$$P(Y = y) = p(y) = e^{-n\lambda}(n\lambda)^y/y!.$$

What does this mean? How is this result related to the results of Exercise 3 on page 176?

8.3. Law of Large Numbers

The law of large numbers is as follows. Suppose $X_1, X_2, \ldots$ is a sequence of independent random variables, and $\overline{X}_n = (X_1 + X_2 + \cdots + X_n)/n$ for any value of n. Further, suppose that

$$EX_i = \mu \quad \text{and} \quad V(X_i) = \sigma^2 \quad \text{for every value of } i$$

and that ε is any given positive number. Then the quantity

$$P\{|\overline{X}_n - \mu| \geq \varepsilon\}$$

tends to zero as n tends to infinity.

What does this mean? It means that the distribution of the variable $\overline{X}_n$ becomes tighter and tighter about the value μ as n gets larger and larger. Note that this is true *whatever the distribution* of the X_i may be.

The proof of the law of large numbers follows from Chebyshev's theorem. Since $E\overline{X}_n = \mu$, $V(\overline{X}_n) = \sigma^2/n$, we can apply the theorem with $\overline{X}_n$ replacing the X of the theorem and $\sigma/n^{1/2}$ replacing the σ of the theorem, to give

$$P\left\{|\overline{X}_n - \mu| \geq h\frac{\sigma}{n^{1/2}}\right\} \leq \frac{1}{h^2}.$$

We now choose our value of h such that, for the given value of ε,

$$h\frac{\sigma}{n^{1/2}} = \varepsilon$$

that is, so that

$$h = \frac{\varepsilon n^{1/2}}{\sigma}$$

whereupon we obtain

$$P\{|\bar{X}_n - \mu| \geq \varepsilon\} \leq \frac{\sigma^2}{n\varepsilon^2}.$$

Now, since σ^2 and ε are fixed, the right-hand side of this inequality becomes smaller and smaller as n get larger and larger, and, in fact, it tends to zero as n tends to infinity. Therefore, the left-hand side must also tend to zero, which proves the law.

Note. One consequence of the law is clear as a practical matter. As we take more and more observations, the average $\bar{X}_n$ becomes more and more "reliable" in the sense that it is less and less variable, that is, is "more like μ." In statistical work when we wish to "estimate" μ by "observing" $\bar{X}_n$, we shall clearly get a more reliable estimate the more observations we take.

Exercises

1. Suppose $X_1, X_2, \ldots$ is a sequence of independent random variables each of which follows the distribution

x	-1	1
$p(x)$	$\frac{1}{2}$	$\frac{1}{2}$

which is such that $EX_i = \mu = 0$, $V(X_i) = \sigma^2 = 1$.
   (a) Find the distribution of $\bar{X}_2 = (X_1 + X_2)/2$.
   (b) Find the distribution of $\bar{X}_4$. [*Hint:* Combine two distributions such as the one in (a).]
   (c) Find the distribution of $\bar{X}_8$. [*Hint:* Combine two distributions such as the one in (b).]
   (d) Find the distribution of $\bar{X}_{16}$. (Similar hint.)
   (e) Plot all these distributions in different figures but using the same abscissa scale and notice how they become more and more concentrated about $\mu = 0$.
2. Repeat Exercise 1 with the distribution

x	-1	1
$p(x)$	$\frac{1}{4}$	$\frac{3}{4}$

3. Repeat Exercise 1 with the distribution

x	-1	0	1
$p(x)$	$\frac{1}{3}$	$\frac{1}{3}$	$\frac{1}{3}$

4. Repeat Exercise 1 with the distribution

x	0	1	2
$p(x)$	$\frac{1}{2}$	$\frac{1}{4}$	$\frac{1}{4}$

WEAKER CONDITIONS. The law of large numbers can also be proved under more general (that is, weaker) conditions, for example, when the X's are uncorrelated rather than independent, or the variances $V(X_i)$ are not all equal. Discussions and comments are to be found in more advanced books. See, for example,

Borel, Émile, *Elements of the Theory of Probability*, translated by John E. Freund, Prentice-Hall, Englewood Cliffs, N.J., 1965.

Feller, William, *An Introduction to Probability Theory and Its Applications*, Vol. I (3rd ed., 1968), Vol. II (1966), Wiley, New York.

Harris, Bernard, *Theory of Probability*, Addison-Wesley, Reading, Mass., 1966.

Parzen, Emanuel, *Modern Probability Theory and Its Applications*, Wiley, New York, 1960.

Révész, Pál, *The Laws of Large Numbers*, Academic Press, New York, 1968.

8.4. Central Limit Theorem

The central limit theorem involves ideas beyond the scope of this book and for this reason we shall not give it the full treatment it really deserves. All we shall do is to mention the idea and emphasize that it is an important one with many practical applications.

We have seen that the law of large numbers says that the distribution of an average $\overline{X}_n$ becomes more and more concentrated about its mean as the number n of components $X_1, X_2, \ldots, X_n$ increases. The central limit theorem says something more about the distribution of a *standardized* average. The result is as follows (in its simplest and least general form).

Suppose $X_1, X_2, \ldots, X_n$ are independent random variables each with the same distribution and such that $EX_i = \mu$, $V(X_i) = \sigma^2$. Then we can obtain a set of standardized independent random variables $Z_1, Z_2, \ldots, Z_n$, by writing

$$Z_i = \frac{X_i - \mu}{\sigma}$$

and so obtain the mean $\bar{Z}_n = (Z_1 + Z_2 + \cdots + Z_n)/n$. Note that

$$\bar{Z}_n = \frac{\bar{X}_n - \mu}{\sigma}.$$

Then, says the central limit theorem, the distribution of $\sqrt{n}\bar{Z}_n$ tends to a standard normal distribution [with mean zero, since obviously $E(\sqrt{n}\bar{Z}_n) = 0$, and with variance one, since $V(\sqrt{n}\bar{Z}_n) = 1$] as n tends to infinity.

A picture of the standard normal distribution is shown in Figure 8.1. It is a continuous distribution, so we shall not discuss it here.

Apart from the conditions specified above, the common distribution followed by the X_i could be of any kind. However, once we have a standardized average $\sqrt{n}\bar{Z}_n$, even when n is quite moderate (say 4 or 5 when the original distribution is roughly symmetrical, or 10 or so otherwise) we can treat $\sqrt{n}\bar{Z}_n$ as though it were a normal variable. Since the standardized normal distribution is well tabulated, we can thus make probability statements about $\sqrt{n}\bar{Z}_n$ even when we do not know the full distribution of the X_i!

We shall mention just one practical application of the central limit theorem (or, actually, of a more general form of it). In industrial situations, observations often vary due to random error. The overall random error in an observation may be a composite of a number of smaller errors such as a meter-reading error, an error in measuring a mixture constituent, an error in setting a temperature control, and so on. Even though we may not know how all the various component errors are distributed, we can, provided the various errors are of comparable magnitude, treat the composite or overall error as being normally distributed by appealing to the central limit theorem. For this reason the central limit theorem and the normal distribution have very

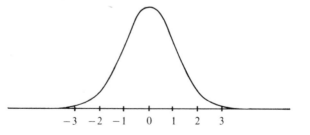

Figure 8.1. Standard normal distribution with mean zero and variance one.

important roles to play in the theory and practical application of statistical techniques.

Exercises

1. State a form of the central limit theorem. Why is it important?
2. Suppose $X_1, X_2, \ldots, X_n$ are independent random variables each of which follows the distribution $p(0) = 1 - p$, $p(1) = p$. What does the central limit theorem tell us in this case? Apply your result to finding the approximate distribution of the number r of successes in n tosses of a coin where the probability of a head on a single toss is p.

CHAPTER 9

Elements of Discrete Markov Processes; Random Walks

9.1. Introduction

You are a gambler and have $90 in your pocket. You enter a certain game and risk $1 at each play. There is a probability p that you will win $1 on each play and a probability $q = 1 - p$ that you will lose your $1 on each play (that is, after the first play you will have $91 or $89, and so on). You wish to build your $90 up to $100, an apparently modest objective, and you will quit if

1. You win that extra $10.
2. You lose your $90 (that is, you are ruined).

What is the probability you will be ruined, what will be your average gain over a number of such games, and how many plays will the game last on the average? The answers will surprise you; they depend on p, of course, as shown in Table 9.1. Clearly, unless p, your chance of winning on each play, exceeds $\frac{1}{2}$, you had best not play!

Table 9.1.   Gambler's Fate in Attempting to Win $10 Using $90

Probability p of a win at each play	Probability of ruin	Average gain ($)	Average length of the game (plays)
$p = 0.50$	0.1	0	900
$p = 0.45$	0.866	-76.6	766
$p = 0.40$	0.983	-88.3	441

Answers such as these are obtained by applying portions of the theory of Markov processes. Some of the elements of this theory will now be given, and a more complete table of results will be derived in Section 9.4, page 199.

A more detailed treatment of the material in Sections 9.1 through 9.4 is given in *Stochastic Processes*, by Emanuel Parzen, published by Holden-Day (San Francisco) in 1962.

9.2. Markov Chains

Suppose that a certain system can be in one of several *states*. Suppose further that, at a discrete set of times, successive observations $X_0, X_1, X_2, \ldots$ are taken. We assume that X_i is a random variable whose value represents the state occupied by the system at the time i, and we shall denote the sequence $X_0, X_1, \ldots, X_i, \ldots$ by the shorthand symbol $\{X_i\}$. If there are only a finite, or countably infinite, number of possible states (see page 80 for the meaning of this), the sequence $\{X_i\}$ is called a *chain*. The sequence $\{X_i\}$ is, more specifically, a *Markov chain* if

1. X_i is a discrete random variable.
2. The probability that X_i takes on any specific value x_i depends only on the most recently observed value, x_{i-1}, and not on any of the X values before it.

Condition 2 implies, for example, that if $x_0, x_1, x_2, \ldots, x_i$ are given,

$$P(X_i = x_i | X_0 = x_0, \ldots, X_{i-1} = x_{i-1}) = P(X_i = x_i | X_{i-1} = x_{i-1}).$$

TRANSITION PROBABILITIES.   Suppose j and k denote two states of the system. To specify the probability law of a Markov chain we need to know the following probabilities:

1. $p_j(i) = P(X_i = j)$, the probability that, at time i, the system is in state j.
2. $p_{jk}(i, l) = P(X_i = k | X_l = j)$, the probability that, at time i, the system is in state k, *given that*, at time l, the system was in state j. These are called the *transition probabilities* of the Markov chain.

9.3. Homogeneous Markov Chains

If $p_{jk}(i, l)$ depends only on the difference $l - i$, the Markov chain is said to be *homogeneous* and in this case a great simplification occurs. We avoid the full derivation of results and simply state what is necessary for this special case. We shall denote states by the integers $1, 2, 3, \ldots$.

ONE-STEP TRANSITION PROBABILITIES. First we need to define what is called a *one-step transition matrix* denoted by the notation $\mathbf{P}_1$ or $\mathbf{P}$ as follows:

$$\mathbf{P}_1 = \mathbf{P} = \begin{array}{c} \text{Next state } k \\ \begin{array}{ccccc} 1 & 2 & 3 & \cdots & k & \cdots \\ \left[\begin{array}{cccccc} p_{11} & p_{12} & p_{13} & \cdots & p_{1k} & \cdots \\ p_{21} & p_{22} & p_{23} & \cdots & p_{2k} & \cdots \\ \vdots & \vdots & \vdots & \vdots & \vdots & \vdots \\ p_{j1} & p_{j2} & p_{j3} & \cdots & p_{jk} & \cdots \\ \vdots & \vdots & \vdots & \vdots & \vdots & \vdots \end{array} \right] & \begin{array}{c} 1 \\ 2 \\ \vdots \\ j \\ \vdots \end{array} \end{array} \end{array} \quad \begin{array}{l} \text{Current} \\ \text{state} \\ j \end{array}$$

The *array* of p_{jk}'s is the *matrix* and is denoted, in short, by the single symbol $\mathbf{P}$. The probability p_{jk} is the probability that, if at one time of observation the system is in state j, it will be in state k at the next time of observation.

Example (Two-State Homogeneous Markov Chain). Suppose a communications system transmits a series of digits which consists entirely of zeros (state 1) and ones (state 2). Suppose each digit transmitted passes through several stages at each of which there is a probability p that the digit is passed through *correctly* and a probability $q = 1 - p$ that the wrong digit emerges. Then the one-step transition matrix for this case is given by $p_{11} = p_{22} = p$, and $p_{12} = p_{21} = 1 - p$, displayed as follows:

Next state

$$\mathbf{P} = \begin{bmatrix} p & 1-p \\ 1-p & p \end{bmatrix}\begin{matrix}1 \\ 2\end{matrix} \Bigg\} \text{Current state}$$

For example, if $p = \frac{3}{4}$ we should have

$$\mathbf{P} = \begin{bmatrix} \frac{3}{4} & \frac{1}{4} \\ \frac{1}{4} & \frac{3}{4} \end{bmatrix}.$$

Now, in some cases it might happen that the probability of transmitting a zero correctly was different from the probability of transmitting a one correctly. For example, we might have

Next

$$\mathbf{P} = \begin{bmatrix} \frac{3}{4} & \frac{1}{4} \\ \frac{1}{3} & \frac{2}{3} \end{bmatrix}\begin{matrix}1 \\ 2\end{matrix} \Bigg\} \text{Current}$$

which means that the chance of transmitting a zero correctly is $\frac{3}{4}$ (and wrongly is $\frac{1}{4}$) while the chance of transmitting a 1 correctly is $\frac{2}{3}$ (and wrongly is $\frac{1}{3}$).

Note that each row sums to one. Clearly this must happen for *any* transition matrix because the elements in any row represent the probabilities that the system will finish in the various states of the system given that it is now in the state designated by the row. Since it is *certain* that *some* state will be next, the total of the row probabilities *must* add to one. In general, then,

$$\sum_k p_{jk} = p_{j1} + p_{j2} + \cdots + p_{jk} + \cdots = 1$$

for all values of j. (*Note carefully:* It is *not* necessarily true that the column probabilities add to one. This did happen in one two-state example above, but this was simply a fluke.)

n-STEP TRANSITION PROBABILITIES.   It can be shown that the n-step transition probabilities for a homogeneous Markov chain can also be written in a matrix array $\mathbf{P}_n$ whose (j, k) element can be denoted by $p_{jk}(n)$.

The elements of $\mathbf{P}_n$ depend on the elements of $\mathbf{P}$ as follows:

$$\mathbf{P}_n = \mathbf{P}^n$$
$$= \mathbf{P} \times \mathbf{P} \times \cdots \times \mathbf{P},$$

there being n matrices in the product. What does this mean? To understand it we have to define the *product of two matrices*. The rule is as follows for square $r \times r$ matrices. Suppose $\mathbf{A}$ and $\mathbf{B}$ are two $r \times r$ matrices as follows:

$$\mathbf{A} = \begin{bmatrix} a_{11} & a_{12} & a_{13} & \cdots & a_{1r} \\ a_{21} & a_{22} & a_{23} & \cdots & a_{2r} \\ a_{31} & a_{32} & a_{33} & \cdots & a_{3r} \\ \cdot & \cdot & \cdot & \cdot & \cdot \\ \cdot & \cdot & \cdot & \cdot & \cdot \\ a_{r1} & a_{r2} & a_{r3} & \cdots & a_{rr} \end{bmatrix},$$

$$\mathbf{B} = \begin{bmatrix} b_{11} & b_{12} & b_{13} & \cdots & b_{1r} \\ b_{21} & b_{22} & b_{23} & \cdots & b_{2r} \\ b_{31} & b_{32} & b_{33} & \cdots & b_{3r} \\ \cdot & \cdot & \cdot & \cdot & \cdot \\ b_{r1} & b_{r2} & b_{r3} & \cdots & b_{rr} \end{bmatrix}.$$

We shall call the product of these matrices $\mathbf{C}$, where $\mathbf{C} = \mathbf{AB}$ and where

$$\mathbf{C} = \begin{bmatrix} c_{11} & c_{12} & c_{13} & \cdots & c_{1r} \\ c_{21} & c_{22} & c_{23} & \cdots & c_{2r} \\ c_{31} & c_{32} & c_{33} & \cdots & c_{3r} \\ \cdot & \cdot & \cdot & \cdot & \cdot \\ c_{r1} & c_{r2} & c_{r3} & \cdots & c_{rr} \end{bmatrix}.$$

The question is: What are the values of the c's in terms of the a's and the b's? To get the c element in the jth row and the kth column, we multiply together, pair by pair, the elements of the jth row of $\mathbf{A}$ with the elements of the kth column of $\mathbf{B}$ and add the results. This sounds more complicated than it actually is to do! For example,

$$c_{12} = a_{11}b_{12} + a_{12}b_{22} + a_{13}b_{32} + \cdots + a_{1r}b_{r2},$$
$$c_{31} = a_{31}b_{11} + a_{32}b_{21} + a_{33}b_{31} + \cdots + a_{3r}b_{r1},$$

and, in general,

$$c_{jk} = a_{j1}b_{1k} + a_{j2}b_{2k} + a_{j3}b_{3k} + \cdots + a_{jr}b_{rk}$$

$$= \sum_{q=1}^{r} a_{jq}b_{qk}.$$

Examples. In the 2×2 case above, where

$$\mathbf{P} = \begin{bmatrix} \frac{3}{4} & \frac{1}{4} \\ \frac{1}{3} & \frac{2}{3} \end{bmatrix},$$

we have

$$\mathbf{P}_2 = \mathbf{P}^2 = \mathbf{P} \times \mathbf{P} = \begin{bmatrix} \frac{3}{4} & \frac{1}{4} \\ \frac{1}{3} & \frac{2}{3} \end{bmatrix} \times \begin{bmatrix} \frac{3}{4} & \frac{1}{4} \\ \frac{1}{3} & \frac{2}{3} \end{bmatrix}$$

$$= \begin{bmatrix} \frac{3}{4}(\frac{3}{4}) + \frac{1}{4}(\frac{1}{3}) & \frac{3}{4}(\frac{1}{4}) + \frac{1}{4}(\frac{2}{3}) \\ \frac{1}{3}(\frac{3}{4}) + \frac{2}{3}(\frac{1}{3}) & \frac{1}{3}(\frac{1}{4}) + \frac{2}{3}(\frac{2}{3}) \end{bmatrix}$$

$$= \begin{bmatrix} \frac{31}{48} & \frac{17}{48} \\ \frac{17}{36} & \frac{19}{36} \end{bmatrix}.$$

As a check, note that the sum of each row is one. The property that the row elements sum to one is true for *all* transition matrices, no matter how many steps are involved. For the transition matrix after three steps we have, now,

$$\mathbf{P}_3 = \mathbf{P}^3 = \mathbf{P}^2 \times \mathbf{P} = \begin{bmatrix} \frac{31}{48} & \frac{17}{48} \\ \frac{17}{36} & \frac{19}{36} \end{bmatrix}\begin{bmatrix} \frac{3}{4} & \frac{1}{4} \\ \frac{1}{3} & \frac{2}{3} \end{bmatrix}$$

$$= \begin{bmatrix} \frac{31}{48}(\frac{3}{4}) + \frac{17}{48}(\frac{1}{3}) & \frac{31}{48}(\frac{1}{4}) + \frac{17}{48}(\frac{2}{3}) \\ \frac{17}{36}(\frac{3}{4}) + \frac{19}{36}(\frac{1}{3}) & \frac{17}{36}(\frac{1}{4}) + \frac{19}{36}(\frac{2}{3}) \end{bmatrix}$$

$$= \begin{bmatrix} \frac{347}{576} & \frac{229}{576} \\ \frac{229}{432} & \frac{203}{432} \end{bmatrix}.$$

Again, as a check, note that each row sums to one, as must happen. We could also get $\mathbf{P}_3$ from the product $\mathbf{P} \times \mathbf{P}^2$ and we should find the same result. Clearly this process could be continued to get $\mathbf{P}_n$ for any specified value of n.

COUNTABLY INFINITE STATES. In this case the **P** matrix is infinite both to the right and downward. Apart from this, all the above steps apply. The matrix multiplication is done in exactly the same way, except that all summations are now infinite.

SUMMARY. The probability system of a homogeneous Markov chain depends only on the one-step transition matrix $\mathbf{P}$. Once this is defined we can get $\mathbf{P}_n$, the n-step transition matrix, for any n by taking the product $\mathbf{P} \times \mathbf{P} \times \cdots \times \mathbf{P} = \mathbf{P}^n$.

Comment. The $\mathbf{P}_n = \mathbf{P}^n$ formula is fairly obvious when one considers what the product achieves. For example, consider $\mathbf{P}_2 = \mathbf{P}^2$. The $(1, 2)$ element of $\mathbf{P}_2$ is given (if there are r states in all) by

$$p_{12}(2) = p_{11}p_{12} + p_{12}p_{22} + p_{13}p_{32} + \cdots + p_{1r}p_{r2}.$$

Now if, at a current stage, the system is in state 1 and, two steps later, the system is in state 2, there are r mutually exclusive ways it could have done this, namely by passing, at the intermediate stage, through states $1, 2, 3, \ldots,$ or r. So if X_1, X_2, and X_3 are the three states at the three stages,

$$
\begin{aligned}
p_{12}(2) = P(X_3 = 2 | X_1 = 1) &= P(X_2 = 1 | X_1 = 1)P(X_3 = 2 | X_2 = 1) \\
&+ P(X_2 = 2 | X_1 = 1)P(X_3 = 2 | X_2 = 2) \\
&+ \cdots \\
&+ P(X_2 = r | X_1 = 1)P(X_3 = 2 | X_2 = r) \\
&= \sum_{k=1}^{r} p_{1k}p_{k2}.
\end{aligned}
$$

as given above. A similar argument (but one more involved symbolically) can be carried out for any number of steps n, as is fairly obvious from the formula in the next paragraph.

THE GENERAL ELEMENT FORMULA. Although the matrix multiplication $\mathbf{P}_n = \mathbf{P}^n$ can be stated *most* simply in matrix form, it can also be stated via summation signs. In general, the (j, k) element of $\mathbf{P}_n$ takes the form

$$p_{jk}(n) = \sum_i \sum_l \cdots \sum_s p_{ji}p_{il}p_{lm} \cdots p_{sk},$$

there being n p's and $(n - 1)$ summations on the right-hand side. Note that the first subscript is j and the last k and, in between, pairs of two consecutive subscripts have the same letter.

It is easy to verify this result, which we give for the record. In practice it is much easier to work with matrices.

Exercises

1. A two-state Markov chain has one-step transition probabilities given by

$$\mathbf{P} = \begin{bmatrix} \frac{1}{2} & \frac{1}{2} \\ \frac{1}{2} & \frac{1}{2} \end{bmatrix}.$$

Find $\mathbf{P}_2$ and $\mathbf{P}_3$. Hence show that $\mathbf{P}_n = \mathbf{P}$ no matter what n may be.

2. A two-state Markov chain has one-step transition probabilities given by

$$\mathbf{P} = \begin{bmatrix} \frac{2}{3} & \frac{1}{3} \\ \frac{1}{3} & \frac{2}{3} \end{bmatrix}.$$

Find $\mathbf{P}_2$, $\mathbf{P}_3$, and $\mathbf{P}_4$ and plot $p_{11}(n)$ against n for $n = 1, 2, 3, 4$. How does it look as if $p_{11}(n)$ will behave as n gets larger and larger? (See Exercise 3 for the general result.)

3. A two-state Markov chain has one-step transition probabilities given by

$$\mathbf{P} = \begin{bmatrix} p & 1 - p \\ 1 - p & p \end{bmatrix}.$$

(a) Find $\mathbf{P}_2$ and $\mathbf{P}_3$ and show that both follow the general formula

$$\mathbf{P}_n = \begin{bmatrix} \frac{1}{2} + \frac{1}{2}(p - q)^n & \frac{1}{2} - \frac{1}{2}(p - q)^n \\ \frac{1}{2} - \frac{1}{2}(p - q)^n & \frac{1}{2} + \frac{1}{2}(p - q)^n \end{bmatrix},$$

where $q = 1 - p$; that is, $p - q = 2p - 1$.

(b) Prove the correctness of the formula given in (a) for $\mathbf{P}_n$ by using induction.[1]

(c) Confirm that the rows of $\mathbf{P}_n$ always add to one.

(d) If the system passes through a large number of steps, in which state is the system most likely to be at the end? Does your answer depend at all on which state the system was in initially?

4. A two-state Markov chain has one-step transition probabilities given by

$$\mathbf{P} = \begin{bmatrix} p_1 & 1 - p_1 \\ 1 - p_2 & p_2 \end{bmatrix}.$$

If $|p_1 + p_2 - 1| < 1$, it can be shown by induction that

$$\mathbf{P}_n = \begin{bmatrix} p_1(n) & 1 - p_1(n) \\ 1 - p_2(n) & p_2(n) \end{bmatrix},$$

where

$$p_1(n) = \frac{1 - p_2 + (1 - p_1)(p_1 + p_2 - 1)^n}{2 - p_1 - p_2},$$

$$p_2(n) = \frac{1 - p_1 + (1 - p_2)(p_1 + p_2 - 1)^n}{2 - p_1 - p_2}.$$

[1] Reminder: (1) Write down the formula with $n - 1$ everywhere in place of n; (2) work out $\mathbf{P}_n = \mathbf{P}\mathbf{P}_{n-1}$ and show that it is of the form given above; (3) confirm that the formula is true for $n = 1$.

(a) Work out $\mathbf{P}_2$ and $\mathbf{P}_3$ and confirm the general formulas for these cases.

(b) Prove the general result using induction. [See Exercise 3(b), footnote.]

5. A two-state Markov chain has one-step transition probabilities given by

$$\mathbf{P} = \begin{bmatrix} 1 & 0 \\ 0 & 1 \end{bmatrix}.$$

After 103 steps, in which state is the system?

6. A four-state Markov chain has one-step transition probabilities given by

$$\mathbf{P} = \begin{bmatrix} 1 & 0 & 0 & 0 \\ \frac{1}{2} & \frac{1}{2} & 0 & 0 \\ \frac{1}{4} & 0 & \frac{1}{2} & \frac{1}{4} \\ \frac{1}{4} & \frac{1}{4} & \frac{1}{4} & \frac{1}{4} \end{bmatrix}.$$

(a) Examine $\mathbf{P}$ and describe how you think the system will behave if it starts in the fourth state.

(b) Evaluate $\mathbf{P}_2$, $\mathbf{P}_3$, and $\mathbf{P}_4$ and see if the conclusion you came to in (a) appears to be confirmed.

7. Repeat (a) and (b) of Exercise 6 when

$$\mathbf{P} = \begin{bmatrix} 0 & 1 & 0 & 0 \\ 0 & 1 & 0 & 0 \\ \frac{1}{3} & \frac{2}{3} & 0 & 0 \\ \frac{1}{3} & \frac{1}{3} & \frac{1}{3} & 0 \end{bmatrix}.$$

8. Repeat (a) and (b) of Exercise 6 when

$$\mathbf{P} = \begin{bmatrix} 1 & 0 & 0 & 0 \\ 0 & 0 & 0 & 1 \\ 0 & 0 & \frac{1}{2} & \frac{1}{2} \\ \frac{1}{2} & \frac{1}{2} & 0 & 0 \end{bmatrix}.$$

9. Repeat (a) and (b) of Exercise 6 when

$$\mathbf{P} = \begin{bmatrix} 1 & 0 & 0 & 0 \\ 1 & 0 & 0 & 0 \\ \frac{1}{4} & \frac{1}{4} & \frac{1}{4} & \frac{1}{4} \\ \frac{1}{4} & \frac{1}{4} & \frac{1}{4} & \frac{1}{4} \end{bmatrix}.$$

10. Repeat (a) and (b) of Exercise 6 when

$$\mathbf{P} = \begin{bmatrix} 0 & 1 & 0 & 0 \\ 0 & 0 & 1 & 0 \\ 0 & 0 & 0 & 1 \\ 1 & 0 & 0 & 0 \end{bmatrix}.$$

11. Repeat (a) and (b) of Exercise 6 when

$$\mathbf{P} = \begin{bmatrix} 0 & 0 & \frac{1}{2} & \frac{1}{2} \\ \frac{1}{4} & \frac{1}{4} & \frac{1}{4} & \frac{1}{4} \\ 1 & 0 & 0 & 0 \\ 1 & 0 & 0 & 0 \end{bmatrix}.$$

AN ABSORBING STATE.   If a state, once attained, cannot be left, it is called an *absorbing* state, since the system is absorbed by it. Clearly the jth state is absorbing if $p_{jj} = 1$.

Examples.   In Exercise 6, 1 is an absorbing state; in 7, 2 is; in 8, 1 is; in 9, 1 is.

Comment.   Although there are a great many additional interesting points to develop for *general* homogeneous Markov chains, we shall not follow them further in this book. Instead we now look in more detail at a particularly interesting type of homogeneous Markov chain called a *random walk*. This type of Markov chain can be used to deal with problems of the type discussed in Section 9.1.

9.4. Random Walks

A random walk is a particular type of Markov chain, with the following property: If the system is in any given state j, then, *in a single transition*, it can only

1. remain in state j, with probability $p_{jj} \geq 0$, or
2. move to one or other of the immediately adjacent states, namely $j - 1$ (with probability $p_{j,j-1} \geq 0$) or $j + 1$ (with probability $p_{j,j+1} \geq 0$).

Of course, necessarily,

$$p_{j,j-1} + p_{jj} + p_{j,j+1} = 1.$$

A one-step transition matrix for a random walk thus takes the form below. (Since any row needs only three p_{jk}'s in it we use three symbols q, r, and p to simplify the subscript notation.) Note that, as we have written it, $\mathbf{P}$ is infinite in all directions, corresponding to an infinite number of states with no boundaries.

Next state

$$\mathbf{P} = \begin{bmatrix} \vdots & \vdots & \vdots & \vdots & \vdots & \vdots & \vdots & \vdots & \vdots \\ \cdots & q_{j-2} & r_{j-2} & p_{j-2} & 0 & 0 & 0 & 0 & \cdots \\ \cdots & 0 & q_{j-1} & r_{j-1} & p_{j-1} & 0 & 0 & 0 & \cdots \\ \cdots & 0 & 0 & q_j & r_j & p_j & 0 & 0 & \cdots \\ \cdots & 0 & 0 & 0 & q_{j+1} & r_{j+1} & p_{j+1} & 0 & \cdots \\ \cdots & 0 & 0 & 0 & 0 & q_{j+2} & r_{j+2} & p_{j+2} & \cdots \\ \vdots & \vdots & \vdots & \vdots & \vdots & \vdots & \vdots & \vdots & \vdots \end{bmatrix} \begin{matrix} \vdots \\ j-2 \\ j-1 \\ j \\ j+1 \\ j+2 \\ \vdots \end{matrix}$$

where the column headers are $\cdots\ j-2\ j-1\ j\ j+1\ j+2\ \cdots$ and the rows are labeled as Current state.

BARRIERS OR BOUNDARIES. A random walk can have barrier or boundary states at one or both ends of its system of states. For example, if we have the states $0, 1, 2, \ldots$, then 0 is a boundary state or a barrier. In this case, the one-step transition probability matrix takes the form

$$\mathbf{P} = \begin{bmatrix} r_0 & p_0 & 0 & 0 & 0 & \cdots \\ q_1 & r_1 & p_1 & 0 & 0 & \cdots \\ 0 & q_2 & r_2 & p_2 & 0 & \cdots \\ 0 & 0 & q_3 & r_3 & p_3 & \cdots \\ \vdots & \vdots & \vdots & \vdots & \vdots & \vdots \end{bmatrix},$$

being infinite to the right and downward. Now we have

$$r_0 + p_0 = 1 \quad \text{and} \quad q_i + r_i + p_i = 1 \quad i = 1, 2, 3, \ldots.$$

Note that:

1. If $r_0 = 1$ and $p_0 = 0$, the barrier at state 0 is *absorbing*; that is, the system never leaves this state once it reaches it. (Example: a gambler who is "ruined" and has no credit.)

2. If $r_0 > 0$ and $p_0 > 0$, the barrier is *reflecting*. (Example: a gambler who is "ruined" but is allowed to play on, his losses being counted only when he has funds to pay.)

When the number of states is finite, there are two boundaries. For example, if there are four states 0, 1, 2, and 3 we have

$$\mathbf{P} = \begin{bmatrix} r_0 & p_0 & 0 & 0 \\ q_1 & r_1 & p_1 & 0 \\ 0 & q_2 & r_2 & p_2 \\ 0 & 0 & q_3 & r_3 \end{bmatrix}.$$

Here

$$r_0 + p_0 = 1 \qquad q_1 + r_1 + p_1 = 1,$$
$$q_3 + r_3 = 1 \qquad q_2 + r_2 + p_2 = 1.$$

States 0 and 3 can each be either absorbing or reflecting according to the values of r_0, p_0, q_3, and r_3.

REASON FOR THE TERM "RANDOM WALK". The term random walk is used because we can regard the variable X_i, the state of the system at time i, as being the position of an object moving on a straight line in such a way that, at each step, the object either

1. remains where it is,
2. moves one step to the right, or
3. moves one step to the left.

As an approximation, random walks can be used to represent diffusing particles, for example in nuclear reactions.

A SPECIAL CASE: THE GAMBLER. Suppose a gambler G is playing a game against a possibly infinitely rich adversary such as a casino. Suppose, on each play, G has a probability p of winning one unit and a probability $q = 1 - p$ of losing one unit. Furthermore, assume that the state 0 is an absorbing barrier; that is, once G loses his fortune (is ruined) play stops. Then $\mathbf{P}$ has a form infinite to the right and downward with $r_0 = 1$, $p_0 = 0$, $r_i = 0$, $p_i = p$, $q_i = 1 - p_i = 1 - p = q$, say, for $i = 1, 2, 3, \ldots$. In other words,

$$
P = \begin{array}{c}
\begin{array}{ccccccc} 0 & 1 & 2 & 3 & 4 & \cdots \end{array} \\
\left[\begin{array}{cccccc}
1 & 0 & 0 & 0 & 0 & \cdots \\
q & 0 & p & 0 & 0 & \cdots \\
0 & q & 0 & p & 0 & \cdots \\
0 & 0 & q & 0 & p & \cdots \\
\vdots & \vdots & \vdots & \vdots & \vdots & \vdots
\end{array}\right]
\begin{array}{c} 0 \\ 1 \\ 2 \\ 3 \\ \vdots \end{array}
\end{array}
$$

where the states are indicated at the top and side of **P**. As we have said, G will stop if he reaches state 0. If he will also stop when his fortune reaches some predetermined value v, then **P** becomes a $(v + 1) \times (v + 1)$ matrix as follows:

$$
P = \begin{array}{c}
\begin{array}{ccccccccc} 0 & 1 & 2 & 3 & \cdots & v-2 & v-1 & v \end{array} \\
\left[\begin{array}{cccccccc}
1 & 0 & 0 & 0 & \cdots & 0 & 0 & 0 \\
q & 0 & p & 0 & \cdots & 0 & 0 & 0 \\
0 & q & 0 & p & \cdots & 0 & 0 & 0 \\
\vdots & \vdots & \vdots & \vdots & \vdots & \vdots & \vdots & \vdots \\
0 & 0 & 0 & 0 & \cdots & q & 0 & p \\
0 & 0 & 0 & 0 & \cdots & 0 & 0 & 1
\end{array}\right]
\begin{array}{c} 0 \\ 1 \\ 2 \\ \vdots \\ v-1 \\ v \end{array}
\end{array}
$$

If we evaluate $\mathbf{P}_n = \mathbf{P}^n$, we can find the various probabilities of getting to any state from any state in n steps.

THE ABSORPTION PROBABILITIES.   What is the probability that, in the finite-state game, G will be absorbed either at 0 or at v? Let $f_{j0}(n)$ be the probability that, starting at state j, G is absorbed at zero at the end of n steps (that is, G loses his fortune at the end of n steps). We can argue that for $j = 0, 1, 2, \ldots, v - 1$,

$$f_{j0}(n) = p_{j0}f_{00}(n - 1) + p_{j1}f_{10}(n - 1) + \cdots + p_{jv}f_{v0}(n - 1),$$

because G will get to 0 after n steps if he goes from j to i after one step and goes from i to 0 in the next $(n - 1)$. Note that we must have

$$f_{00}(m) = 1 \qquad f_{v0}(m) = 0$$

for any value of m. If we imagine n to be very large, we can take it to be approximately true that

$$f_{j0}(n) = f_{j0}(n - 1)$$

and call both these simply f_{j0}. (Roughly speaking, one step more or less in a large number of steps cannot affect the probabilities much!) This means that we can write

$$f_{j0} = p_{j0}f_{00} + p_{j1}f_{10} + \cdots + p_{jv}f_{v0}$$

with

$$f_{00} = 1 \qquad f_{v0} = 0.$$

Since

$$p_{j,j+1} = p \qquad p_{j,j-1} = q = 1 - p$$

and

$$p_{jk} = 0 \quad \text{if } k \neq j - 1 \quad \text{or} \quad j + 1,$$

we get the *recurrence equation*

$$f_{j0} = qf_{j-1,0} + pf_{j+1,0},$$

subject to *boundary conditions* $f_{00} = 1$, $f_{v0} = 0$. To get the general solution of a recurrence equation is not hard but is beyond the scope of this book. We shall simply state the solution and ask the reader to verify that it does satisfy the recurrence equation. The general solution has two separate cases as follows:

1. $f_{j0} = \dfrac{(q/p)^j - (q/p)^v}{1 - (q/p)^v}$ for $p \neq q$.

2. $f_{j0} = \dfrac{v - j}{v}$ for $p = q = \frac{1}{2}$.

VERIFICATIONS. To verify that these solutions are valid, substitute them into the recurrence equation. Note also that the boundary conditions are satisfied; that is, $f_{00} = 1$, $f_{v0} = 0$, in both cases.

The probability, f_{jv}, that, starting at state j, G will be absorbed at v is simply $1 - $ (the probability of absorption at 0), because, when absorption does take place, it can be only at 0 or v. Thus

$$f_{jv} = 1 - f_{j0}.$$

THE EXPECTED GAIN.   If G played a large number of games, starting each time with a fortune j and quitting when he reached 0 or v, what would he expect to gain on the average? Since he will gain

$$g = v - j \qquad \text{with probability } 1 - f_{j0}$$

and

$$g = -j \qquad \text{with probability } f_{j0},$$

his expected gain is

$$\begin{aligned}
Eg &= (v - j)(1 - f_{j0}) - jf_{j0} \\
&= v(1 - f_{j0}) - j \\
&= v - j - vf_{j0}.
\end{aligned}$$

This is negative if $p < q$, positive if $p > q$, and zero if $p = q$.

HOW LONG WILL THE GAME LAST, ON THE AVERAGE?   By applying appropriate Markov chain theory it is possible to find the *mean time to absorption*, which we shall denote by MTTA, that is, the average time, over a large number of games, before G is absorbed at 0 or v. We give the following results without proof and ask the reader to accept them. Again the results are different for $p \neq q$ and for $p = q = \frac{1}{2}$:

1. $\displaystyle \text{MTTA} = \frac{1}{q - p}\left\{ j - v\frac{1 - (q/p)^j}{1 - (q/p)^v} \right\}$

$\displaystyle \qquad\quad = \frac{1}{q - p}\{ j - v(1 - f_{j0}) \}$

$\displaystyle \qquad\quad = \frac{Eg}{p - q} \qquad \text{for } p \neq q.$

Note that this is *always* positive, because $Eg = v(1 - f_{j0}) - j$ and $p - q$ always have the same sign.

2. $\text{MTTA} = j(v - j)$,

$\qquad\qquad = jvf_{j0}, \qquad \text{for } p = q = \frac{1}{2}.$

The j in the above formulas is, of course, G's *initial* state.

Example. G enters a game in which he has probability $p = 0.45$ of winning one unit and probability $q = 1 - p = 0.55$ of losing one unit at each turn. If G starts the game with $j = 90$ units and he would like to attain $v = 100$ units, what is the probability $f_{90,0}$ that he will be ruined, what will be his average gain Eg, and how long would he expect the game to last, that is, what is the MTTA?

We substitute in the various formulas above. Since $p \neq q$ and $q/p = \frac{11}{9}$,

$$f_{90,0} = \frac{(\frac{11}{9})^{90} - (\frac{11}{9})^{100}}{1 - (\frac{11}{9})^{100}}$$

Now $(\frac{11}{9})^{100}$ is enormously large compared with 1. Thus we can ignore the 1 in the denominator and get, as a very close approximation to $f_{90,0}$, the value

$$\frac{(\frac{11}{9})^{90} - (\frac{11}{9})^{100}}{-(\frac{11}{9})^{100}} = 1 - \left(\frac{9}{11}\right)^{10} = 0.8655.$$

It follows immediately that

$$Eg = 10 - 100(0.8655) = -76.55$$

and that

$$\text{MTTA} = \frac{-76.55}{0.45 - 0.55} = 765.5.$$

SOME RESULTS FOR VARIOUS SITUATIONS. Table 9.2 shows some results of the calculations above for various values of p, j, and v. It will be seen that Table 9.1 in Section 9.1 was extracted from the results given here.

DISCUSSION OF TABLE 9.2. The first 10 entries in Table 9.2 show how the game possibilities change as the initial fortune j is changed for $p = 0.49$, that is, when the odds are only *slightly* against the gambler G. Clearly the more G wishes to win, the greater is the probability of his being ruined. However, since p is so close to $\frac{1}{2}$, G has (except when $j = 5, v = 10$) a better than 50 % chance of winning. His expected or average gain is negative, however; while he loses less than 50 % of the time on the average (except for $j = 5, v = 10$), he loses more when he loses than he wins when he wins.

Some other very interesting comparisons can be made from the lower part of Table 9.2. For example, suppose $p = 0.30$ and our gambler G wants to win one additional unit. If he goes into the game with $j = 9$ units, his chance of success is $1 - f_{90} = 1 - 0.572 = 0.428$. However, his chance of success if $j = 99$ or $j = 999$ is $1 - 0.571 = 0.429$, that is, only 0.001 more! Now look at his expected or average gain; it is -4.7 if $j = 9$, -56.1 if $j = 99$, and -570.4 if $j = 999$! Moreover, the larger j is, the more he actually *can* lose in one play of the game. Let us also look at the case where $p = 0.45$; the probability of winning one unit is now $1 - 0.210 = 0.790$ when $j = 9$, and rises only slightly to 0.818 when $j = 99$ or 999. The respective average gains are -1.1, -17.1, and -180.8 and, again, the larger j is, the more G actually *can* lose in a single game.

Table 9.2.   A Gambler's Fate in Various Situations

Initial fortune j	Target value v	Probability p of a win at each play	Probability of ruin	Average gain	Average length of the game (plays)
5	10	0.49	0.550	−0.5	25
6	10	0.49	0.448	−0.5	24
7	10	0.49	0.343	−0.4	22
8	10	0.49	0.233	−0.3	17
9	10	0.49	0.119	−0.2	9
91	100	0.49	0.308	−21.8	1090
93	100	0.49	0.249	−17.9	894
95	100	0.49	0.185	−13.5	673
97	100	0.49	0.115	−8.5	426
99	100	0.49	0.040	−3.0	150
9	10	0.50	0.1	0	9
9	10	0.45	0.210	−1.1	11
9	10	0.40	0.339	−2.4	12
9	10	0.35	0.462	−3.6	12
9	10	0.30	0.572	−4.7	12
90	100	0.50	0.1	0	900
90	100	0.45	0.866	−76.6	766
90	100	0.40	0.983	−88.3	441
90	100	0.35	0.998	−89.8	299
90	100	0.30	0.9998	−90.0	225
99	100	0.50	0.01	0	99
99	100	0.45	0.182	−17.1	172
99	100	0.40	0.333	−32.3	162
99	100	0.35	0.462	−45·2	151
99	100	0.30	0.571	−56.1	140
999	1000	0.50	0.001	0	999
999	1000	0.45	0.182	−180.8	1808
999	1000	0.40	0.333	−332.3	1662
999	1000	0.35	0.462	−460.5	1535
999	1000	0.30	0.571	−570.4	1426

The overall practical moral for a game of this type is clear. When the odds are against you, that is, $p < \frac{1}{2}$, you will, in any case, lose in the long run and you will lose on the average more, the more you risk. Therefore, *risk only what you can afford to lose*. Although you may win, the odds are against it.

Of course, you may feel it worthwhile to pay for your pleasure in playing. Suppose you wish to win one unit. Since the average lengths of the games for $j = 9, 99$, and 999 are 12, 140, and 1426 for $p = 0.30$ and 11, 172, and 1808 for $p = 0.45$, you will obtain a longer game on the average (and so perhaps more pleasure?) by using a larger initial capital; however, you will lose more on the average. Only the individual concerned can measure his personal pleasure in playing and compare it with his chagrin at his inevitable average loss when $p < \frac{1}{2}$. Our advice remains: Risk only what you can afford to lose!

Exercises

1. A random walk has a one-step transition matrix of the form

$$
\begin{array}{cccc}
 & 0 & 1 & 2 \\
\mathbf{P} = & \begin{bmatrix} 1 & 0 & 0 \\ q & 0 & p \\ 0 & 0 & 1 \end{bmatrix} & \begin{matrix} 0 \\ 1 \\ 2 \end{matrix}
\end{array}
$$

(a) Use the f_{j0} formula to find f_{10}.
(b) Why is this answer obvious from an inspection of $\mathbf{P}$?
(c) What are the expected gain and the expected duration of a "game," beginning with $j = 1$?

2. A random walk has a one-step transition matrix of the form (with $p \neq q$)

$$
\begin{array}{ccccc}
 & 0 & 1 & 2 & 3 \\
\mathbf{P} = & \begin{bmatrix} 1 & 0 & 0 & 0 \\ q & 0 & p & 0 \\ 0 & q & 0 & p \\ 0 & 0 & 0 & 1 \end{bmatrix} & \begin{matrix} 0 \\ 1 \\ 2 \\ 3 \end{matrix}
\end{array}
$$

(a) Find f_{10}, f_{20}. [Answers: $f_{10} = q/(p^2 + q)$; $f_{20} = q^2/(p^2 + q)$.]
(b) What are the expected gain and the expected duration of a "game" beginning (1) with $j = 1$, (2) with $j = 2$?

3. Using the formulas given in this section, obtain any five lines of Table 9.2. (Select lines from various sections of the table.)

4. Make up four counters of one color (we shall call it green here) and six counters of another color (red) and mix them up. Imagine you have a fortune of nine units to bet with. Make a random drawing of one counter and add one unit to your fortune if the drawn counter is green ($p = 0.40$) or take one unit away from your fortune if the drawn counter is red

$(1 - p = 0.60.)$ Replace the counter, mix, redraw, and continue in this way until you have zero or 10 units in hand, when the game is ended. Play a succession of games, keeping a game record as follows:

Game No.	Ruin	Achieved 10	Gain	Plays per game
1	1	0	-9	17
2	0	1	1	3
3	0	1	1	13
4	1	0	-9	21

Find the average of each column and compare these averages with the $j = 9$, $v = 10$, $p = 0.40$ entries in Table 9.2. (*Note:* The random mechanism could also consist of a bag containing 4 "green" marbles and 6 "red" ones, where "green" and "red" simply stand for whatever colors you have available.)

9.5. One-Step Equiprobable Random Walks from the Origin, without Barriers

We now mention a special case of the infinite-state type of random walk introduced at the beginning of Section 9.4. Suppose the walk starts from 0 and can proceed, in unit steps, in either direction, to plus or minus infinity. Furthermore, suppose $p_j = \frac{1}{2}$, $q_j = 1 - p_j = \frac{1}{2}$, $r_j = 0$ for all values of $j = 0, \pm 1, \pm 2, \ldots$. Then we have a one-step transition matrix as follows:

$$
P =
\begin{array}{c}
\begin{array}{ccccccccc}
\cdots & -3 & -2 & -1 & 0 & 1 & 2 & 3 & \cdots
\end{array} \\
\left[
\begin{array}{ccccccccc}
\vdots & \vdots & \vdots & \vdots & \vdots & \vdots & \vdots & \vdots & \vdots \\
\cdots & \frac{1}{2} & 0 & \frac{1}{2} & 0 & 0 & 0 & 0 & \cdots \\
\cdots & 0 & \frac{1}{2} & 0 & \frac{1}{2} & 0 & 0 & 0 & \cdots \\
\cdots & 0 & 0 & \frac{1}{2} & 0 & \frac{1}{2} & 0 & 0 & \cdots \\
\cdots & 0 & 0 & 0 & \frac{1}{2} & 0 & \frac{1}{2} & 0 & \cdots \\
\cdots & 0 & 0 & 0 & 0 & \frac{1}{2} & 0 & \frac{1}{2} & \cdots \\
\vdots & \vdots & \vdots & \vdots & \vdots & \vdots & \vdots & \vdots & \vdots
\end{array}
\right]
\begin{array}{c}
\vdots \\
-2 \\
-1 \\
0 \\
1 \\
2 \\
\vdots
\end{array}
\end{array}
$$

In fact, however, this particular random walk is best handled without matrices; a complete treatment is given by William Feller in Chapter 3 of *An Introduction to Probability Theory and Its Applications*, Vol. I, Wiley, New York, 3rd ed., 1968. In this book we shall not go into the mathematical details but will present some of the important formulas.

The simplest application of this type of random walk is to coin tossing. If A and B toss a fair coin and A wins one unit for a head and loses one unit for a tail, then the state of A's fortune compared with his initial fortune is given by a random walk of this type (and similarly for B, of course). The state 0 is A's initial state and he will be "up" (on the winning or positive side), "down" (on the losing or negative side), or "even" (back at 0) throughout the game. We can see that various quantities might be of interest: The state A is in; how far to the positive side he may reach; when he last was "even" in a sequence of results; the chance that he will first again be "even" at some given step; and so on. We shall give below, without proof, some of the important results for this problem.

DEFINITION OF VARIABLES.   Let the random variable sequence $X_0 = 0, X_1, X_2, \ldots, X_i, \ldots$ denote the states reached by successive steps of the random walk, starting at $X_0 = 0$. Let

$$M_i = \text{maximum of } (X_0, X_1, X_2, \ldots, X_i)$$

denote the highest value X achieves in i steps, that is, the farthest distance the particles gets to the right of the origin in i steps. Further, let N_i (which can take as its value any even integer less than or equal to i) denote the step number at which the particle is *last* at the origin during the finite sequence of steps $0, 1, 2, \ldots, i$. Then X_i, M_i, and N_i are all random variables and have discrete distributions which can be evaluated as follows. (The symbol j denotes a state as usual.)

THE DISTRIBUTIONS OF X_i, M_i, AND N_i

1. $P(X_i = j) = \begin{pmatrix} i \\ \frac{1}{2}(i+j) \end{pmatrix} (\frac{1}{2})^i \qquad \text{for } j = -i, -i+2, \ldots, i-2, i.$

For other values of j, $P(X_i = j) = 0$ since, because of the way the walk takes place, these other values of j cannot be reached in i steps. (For example, if $i = 2$, we cannot get to $j = 1$ in two steps; see Figure 9.1.)

An idea of the general shape of this distribution can be obtained by considering the case $i = 4$ when we have

j	-4	-2	0	2	4
$P(X_4 = j)$	$\frac{1}{16}$	$\frac{4}{16}$	$\frac{6}{16}$	$\frac{4}{16}$	$\frac{1}{16}$

How this distribution arises can be seen from Figure 9.1, which shows all the possible 16 paths of a four-step random walk. Here the numbers at the points are states and the numbers on the lines indicate the number of possible paths that use the segment indicated. Clearly there is only 1 way, out of 16, to finish at -4; there are 4 ways out of 16 to finish at -2; and so on.

2. $P(M_i = j) = P(X_i = j) + P(X_i = j + 1)$ for $j = 0, 1, 2, \ldots, i$.

[*Note:* Since $P(X_i = j)$ is defined only for *every other integer* between $-i$ and i, one of these two terms is always zero.]

An idea of the general shape of this distribution can be obtained by considering the case $i = 4$ when we have

j	0	1	2	3	4
$P(M_4 = j)$	$\frac{6}{16}$	$\frac{4}{16}$	$\frac{4}{16}$	$\frac{1}{16}$	$\frac{1}{16}$

How this distribution arises can be seen from Figure 9.1 by counting, for each j, the distinct paths that go no farther than (but *do go as far as*) j. For example, for $j = 1$, the four paths

X_0	X_1	X_2	X_3	X_4
0	1	0	1	0
0	1	0	-1	-2
0	1	0	-1	0
0	-1	0	1	0

are the only ones that attain but do not go beyond $j = 1$, and so on.

3. $P(N_i = u) = \begin{pmatrix} u \\ \frac{1}{2}u \end{pmatrix} \begin{pmatrix} e(i) - u \\ \frac{1}{2}[e(i) - u] \end{pmatrix} (\tfrac{1}{2})^{e(i)}$ for $u = 0, 2, 4, \ldots, e(i)$,

where $e(i)$ is the largest even integer contained in i. In other words, $e(i) = i$ if i is even, or $e(i) = i - 1$ if i is odd. This is a "U-shaped" distribution, higher at the ends than in the middle. It is symmetrical about $\frac{1}{2}e(i)$. To see this in general we can write u as $e(i) - [e(i) - u]$, and interchange the first and second factors above. The foregoing probability then becomes

$$P(N_i = u) = \begin{pmatrix} e(i) - u \\ \frac{1}{2}[e(i) - u] \end{pmatrix} \begin{pmatrix} e(i) - [e(i) - u] \\ \frac{1}{2}e(i) - \frac{1}{2}[e(i) - u] \end{pmatrix} (\tfrac{1}{2})^{e(i)}$$

$$= P[N_i = e(i) - u].$$

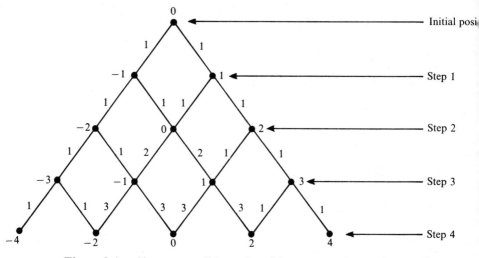

Figure 9.1. Sixteen possible paths of four steps of a random walk.

For the case $i = 4$ we obtain the following results:

u	0	2	4
$P(N_4 = u)$	$\frac{6}{16}$	$\frac{4}{16}$	$\frac{6}{16}$

How this distribution arises can be seen from Figure 9.1. Of the 16 possible paths of a four-step random walk, six (one finishing at -4, two finishing at -2, two finishing at 2, one finishing at 4) never return to the origin after they originally leave it. Thus we get $P(N_4 = 0) = \frac{6}{16}$ for the $u = 0$ case. Also, four paths return to zero at step 2 but then finish (two each) at -2 or 2. Thus $P(N_4 = 2) = \frac{4}{16}$. Again, six paths finish at 0 on the fourth step. Thus $P(N_4 = 4) = \frac{6}{16}$.

PROBABILITIES OF VARIOUS EVENTS. The probability that a *first* return to the origin 0 takes place at the kth step (k even, of course) is

4. $f_k = \dfrac{1}{k-1} \binom{k}{\frac{1}{2}k} (\tfrac{1}{2})^k.$

The probability that a return to 0 takes place at the second step ($k = 2$) is thus

$$f_2 = \binom{2}{1}\left(\frac{1}{2}\right)^2 = \frac{1}{2}.$$

Clearly this is so because the possible paths are

X_0	X_1	X_2
0	1	2
0	1	0
0	−1	0
0	−1	−2

and exactly half of the four possible equiprobable paths return to zero. When $k = 4$,

$$f_4 = \frac{1}{3}\binom{4}{2}\left(\frac{1}{2}\right)^4 = \frac{1}{8},$$

as can easily be confirmed by enumerating all 16 equiprobable possible paths, as in Figure 9.1, and seeing that only two, namely

X_0	X_1	X_2	X_3	X_4
0	1	2	1	0
0	−1	−2	−1	0

return to the origin *for the first time* at the fourth step. (Four other paths return for the *second* time to the origin at the fourth step. This is easily seen from Figure 9.1, where, as we have mentioned, the numbers on the segments indicate the number of possible paths that contain the segments indicated.)

The probability that the first passage through r occurs at the kth step (that is, that the process "hits" state r for the first time at the kth step) is

5. $\phi_{rk} = \dfrac{r}{k}\left(\dfrac{k}{\frac{1}{2}(k + r)}\right)(\tfrac{1}{2})^k,$

where it is understood that $\phi_{rk} = 0$ if $k + r$ is odd. For example, if $r = 2$, $k = 3$,

$$\phi_{23} = \frac{2}{3}\binom{3}{\frac{5}{2}}\left(\frac{1}{2}\right)^3 = 0,$$

since $\binom{3}{\frac{5}{2}}$ is not defined. However if $r = 1$, $k = 3$,

$$\phi_{13} = \frac{1}{3}\binom{3}{2}\left(\frac{1}{2}\right)^3 = \frac{1}{8}.$$

This is obvious from Figure 9.1, where only one of the eight possible paths at step 3 reaches $r = 2$, *for the first time* at that step.

The probability that the rth return to the origin occurs at step k is given by

6. $\phi_{r,k-r} = \dfrac{r}{k-r}\binom{k-r}{\frac{1}{2}k}(\frac{1}{2})^{k-r}.$

For example, if $r = 2$, $k = 4$,

$$\phi_{22} = \frac{2}{2}\binom{2}{2}\left(\frac{1}{2}\right)^2 = \frac{1}{4}.$$

This is obviously true from Figure 9.1, where four of the 16 possible equi-probable paths at step 4 ($k = 4$) return to the origin for the second time ($r = 2$).

IMPLICATIONS. The theory of these random walks has some remarkable implications, remarkable in the sense that they seem to be contrary to intuition and common sense. For the full flavor of these apparent contradictions, Feller's Chapter 3 should be read in detail. Here we shall merely make one of his points to whet the reader's appetite. All quotations are from Feller's book.

Intuitively we might feel "that in a long coin-tossing game each player will be on the winning side for about half the time and that the lead will pass not infrequently from one player to the other." However, the U-shaped symmetrical distribution of N_i (formula 3) implies that "*with probability $\frac{1}{2}$, no equalization occurred in the second half of the game regardless of the length of the game.* Furthermore, the probabilities near the end points are greatest; the most probable values are the extremes 0 and i. The following numerical result shows how unintuitive actual coin-tossing behavior can be:

"*Example (a).* Suppose that a great many coin-tossing games are conducted simultaneously at the rate of one per second, day and night, for a whole year. On the average, in one out of ten games the last equalization will occur before 9 days have passed, and the lead will not change during the following 356 days. In one out of twenty cases the last equalization takes place within $2\frac{1}{4}$ days, and in one out of a hundred cases it occurs within the first 2 hours and 10 minutes."

Another possible consequence of this type of phenomenon is contained in the following example.

"*Example (b).* Suppose that in a learning experiment lasting one year a child was consistently lagging except, perhaps, during the initial week. Another child was consistently ahead except, perhaps, during the last week. Would the two children be judged equal? Yet, let a group of 11 children be exposed to a similar learning experiment involving no intelligence but only chance. One among the 11 would appear as leader for all but one week, another as laggard for all but one week."

Feller reports the results of a particular computer-simulated coin tossing record of 10,000 "tosses" in which, "starting from the origin

the path stays on the

negative side	positive side
for the first 7804 steps	next 8 steps
next 2 steps	next 54 steps
next 30 steps	next 2 steps
next . 48 steps	next 6 steps
next 2046 steps	
Total of 9930 steps	Total of 70 steps
Fraction of time: 0.993	Fraction of time: 0.007

"This *looks* absurd, and yet the probability that in 10,000 tosses of a perfect coin the lead is at one side for more than 9930 trials and at the other for fewer than 70 exceeds $\frac{1}{10}$. In other words, on the average *one record out of ten will look worse than the one just described*. By contrast, the probability of a balance better than [that shown] is only 0.072.

"... Sampling of expert opinion revealed that ... nobody counted on the possibility of only 8 changes of sign. Actually the probability of not more than 8 changes of sign exceeds 0.14.... If [the results] seem startling, this is due to our faulty intuition and to our having been exposed to too many vague references to a mysterious 'law of averages'."

Exercises

1. Use formulas 1, 2, and 3 of Section 9.5 to find the exact distributions of X_i, M_i, and N_i for $i = 5$. Check that the probabilities add to one in every case. Draw a figure like Figure 9.1 to check your results, and draw diagrams of the three distributions.
2. Use formula 4 of Section 9.5 to find $f_2, f_4, f_6, f_8, f_{10}$. Does it look as if the sum $f_2 + f_4 + f_6 + \cdots$ will converge, and if so to what sum?
3. Use formula 5 of Section 9.5 to find the probability ϕ_{rk} that the first passage through r occurs at step k for $k = 1, 2, 3, 4, 5$, and for $r = 1$ and 2.
4. Use formula 6 of Section 9.5 to find the probabilities that the first, second, third, fourth, and fifth returns to the origin will occur at the tenth step. What are the conditional probabilities that the second to fifth returns to the origin will occur at the tenth step, *given* that the first return takes place at step 2? Why are your answers obvious?

9.6. Suggested Additional Reading

Feller, William, *An Introduction to Probability Theory and Its Applications,* Vol. I, Wiley, New York, 3rd ed., 1968. (Also Vol. II, 1966.)

Karlin, Samuel, *A First Course in Stochastic Processes,* Academic Press, New York, 1966.

Parzen, Emanuel, *Stochastic Processes,* Holden-Day, San Francisco, 1962.

CHAPTER 10

Elementary Decision-Making
Using Expectations

10.1. Introduction

Interest in probability originally developed from games of chance, and many of the "straight" applications of probability occur in that area. This we have seen demonstrated via the card and dice examples that run through this book, and also in the gambling applications in Chapter 9. However, in recent years, probability theory has also been used in business and industrial situations, where decisions have to be made between several alternatives, and where there is uncertainty as to which choice will give rise to the greatest advantage. (By "advantage" one could mean many things, for example, profit, flexibility, growth, and so on.) In this short chapter, we provide a few simple examples of probabilistic decision-making, based on the use of expected values. Much more sophisticated work is possible, as the many advanced courses in business schools all over the world testify. However, the subject of decision-making is not the main purpose of this book; for this reason we merely touch upon the topic in the most elementary manner and hope that the reader will (if he is interested) find out more in other courses or by further reading.

10.2. Expectation: A Recapitulation

As we intimated above, a basic concept in decision-making is that of *mathematical expectation,* or *expected value,* defined in Chapter 5. To review, if the probability that a discrete random variable X attains the value x is given by $p(x)$, then the mathematical expectation of the random variable is defined to be

$$EX = \sum_x xp(x),$$

where the summation is over all values of x.

Example 1.   Suppose the random variable X is the number showing on a tossed die. Then the mathematical expectation of the result of a single toss is

$$EX = 1(\tfrac{1}{6}) + 2(\tfrac{1}{6}) + 3(\tfrac{1}{6}) + 4(\tfrac{1}{6}) + 5(\tfrac{1}{6}) + 6(\tfrac{1}{6})$$
$$= 3\tfrac{1}{2}.$$

Of course, we never actually get $3\tfrac{1}{2}$ when we toss a die, even though the distribution has this mean value. However, suppose we were to make a very large number of tosses n and were to evaluate the mean score $\overline{X}_n$. Since $E\overline{X}_n = 3\tfrac{1}{2}$ also, and since, by the law of large numbers (Section 8.3), the distribution of $\overline{X}_n$ is more concentrated around $3\tfrac{1}{2}$ than is the distribution of X, we should expect $\overline{X}_n$ to be quite close to $3\tfrac{1}{2}$. (There is a very small probability that $\overline{X}_n$ will not be close to $3\tfrac{1}{2}$, but this is quite unlikely if n is large.)

Example 2.   Suppose that the probability that a house of a particular construction located in a certain neighborhood and insured for \$15,000 will be destroyed by fire during the 3-year term of the insurance is 0.001. If the insurance company charges a premium of \$25 for each such policy, what is the expected profit per policy? (We shall assume that a house is either untouched or is destroyed completely.) We take as the random variable X the amount of profit received by the insurance company. Thus X has the value \$25 with probability $1 - 0.001 = 0.999$, which is the probability that the house is untouched, and X has the value \$25 − \$15,000 = − \$14,975 with probability 0.001. Thus the expected profit per policy is given by

$$\$25(0.999) + (-\$14,975)(0.001) = \$10,$$

and (see Example 1) the company would expect to gain \$10 per policy on the average over a large number of policies.

Example 3.   Suppose that two players, A and B, play a game of cards as follows. Player A draws a card from an ordinary bridge deck; he receives from B \$5 for a spade, \$4 for a heart, \$3 for a diamond, and \$2 for a club. However, although still receiving these amounts, he pays to B \$8 for a queen, king, or ace and \$5 for a two, three, or four. This is an example of a simple statistical game with two players and with payoffs from one to another: a "two-person, zero-sum" game, meaning that the amount that one of two players loses, the other player wins. Let the random variable X denote the amount *won* by A on each card drawn. Suppose the card drawn is a spade. Then A wins

$0 for a two, three, or four, with probability $\frac{3}{52}$,

$5 for a five to jack, with probability $\frac{7}{52}$,

− \$3 for a queen, king, or ace, with probability $\frac{3}{52}$,

For a heart we get corresponding figures of

$$-\$1 \qquad \tfrac{3}{52}$$
$$\$4 \qquad \tfrac{7}{52}$$
$$-\$4 \qquad \tfrac{3}{52}$$

For a diamond the corresponding figures are

$$-\$2 \qquad \tfrac{3}{52}$$
$$\$3 \qquad \tfrac{7}{52}$$
$$-\$5 \qquad \tfrac{3}{52}$$

Finally, for a club, the corresponding figures are

$$-\$3 \qquad \tfrac{3}{52}$$
$$\$2 \qquad \tfrac{7}{52}$$
$$-\$6 \qquad \tfrac{3}{52}$$

It follows that, in dollars,

$$EX = 0(\tfrac{3}{52}) + 5(\tfrac{7}{52}) - 3(\tfrac{3}{52}) - \cdots - 6(\tfrac{3}{52})$$
$$= \tfrac{26}{52}$$
$$= 0.50.$$

B's expectation is obviously -0.50. The interpretation (See Example 1) is that, in a long series of games, A will expect to win an average of about \$0.50 per game, while B will expect to lose the same amount.

An alternative and simpler method of handling this particular example is to treat A's gains and losses as two entirely separate games. Thus, for the "gains" game, A wins \$5 with probability $\frac{13}{52} = \frac{1}{4}$ for a spade, wins \$4 with probability $\frac{1}{4}$ for a heart, wins \$3 with probability $\frac{1}{4}$ for a diamond, and wins \$2 with probability $\frac{1}{4}$ for a club. His "gain" expectation is thus

$$EG = 5(\tfrac{1}{4}) + 4(\tfrac{1}{4}) + 3(\tfrac{1}{4}) + 2(\tfrac{1}{4}) = 3.50.$$

His "loss" expectation is similarly

$$EL = 8(\tfrac{3}{13}) + 5(\tfrac{3}{13}) + 0(\tfrac{7}{13}) = 3.00.$$

Thus his overall expectation is

$$EG - EL = 3.50 - 3.00 = 0.50,$$

exactly as before.

Exercises

1. You enter a lottery with 299 other persons.
   - (a) Each person pays \$1 and a name is selected at random from the 300 names to win the \$300 pot. What is your mathematical expectation?
   - (b) One of the 300 names is selected at random and the "lucky" person has to pay each of the others \$1. What is your mathematical expectation?
   - (c) Compare the expectations in (a) and (b). In which lottery would you prefer to take part? Why?
2. If one ball is drawn from a box containing 30 red, 20 blue, and 50 white balls, and if you win \$1 for red, \$2 for blue, but lose \$3 for white, find your mathematical expectation.
3. You and a friend are tossing coins. You toss a cent, a nickel, a dime, and a quarter together. You receive from your friend the amount of each coin showing heads and you pay him 10 cents for each coin showing tails. What is your expected net gain (or loss) per game?
4. A game is played as follows. Person A pays person B \$4. B then tosses a single fair die a maximum of four times; if, on any of the four possible tosses, a 1 or 6 turns up, the game terminates and B pays A \$5. Otherwise A simply loses the initial \$4. What gain can A expect on the average?
5. Lloyds of London figures that a certain loss will occur only once in 10,000 cases. If a loss occurs, Lloyds will have to pay \$25,000. What premium should Lloyds charge for insuring against such a loss in order to break even in the long run? What premium should Lloyds charge to make an average of \$5 per policy? What is Lloyds' expected profit per policy if the premium is set at \$3.25?

6. A certain gambling casino pays the following odds for the total score resulting from a single toss of two dice:

2 and 12 each pay 30 to 1	5 and 9 each pay 6 to 1
3 and 11 each pay 15 to 1	6 and 8 each pay 5 to 1
4 and 10 each pay 9 to 1	7 pays 4 to 1

Assuming the dice are balanced, what is the expected net gain (or loss) per game for a player who places a $1 bet on each of the 11 numbers simultaneously? Which number is the best bet for a player?

10.3. Making Decisions under Uncertainty

The application of the concept of mathematical expectation to *making decisions under uncertainty* will be illustrated by the following examples.

Example 1.   Suppose that Jones is in charge of arranging a benefit for a charitable organization and that he has narrowed down his list of possibilities to two: a baseball game at the stadium or a movie at a large theater, both for the same day. That is, he can arrange for two major league teams to play an exhibition ball game or he can arrange for a special showing of a new movie, with some of the proceeds going to his organization in either case. He is now faced with the task of making a decision when there is some uncertainty about the outcome.

If he arranges a ball game and the weather is fine, there will be a good turnout and the expected net proceeds will be about $7000. (This is a reasonable estimate based upon the experience of others.) If the weather is not fine, the expected net proceeds will be about $1000. On the other hand, if the weather is fine the movie will net $3000, but if the weather is not fine, people will be looking for indoor entertainment and the movie will net $6000. Which is the rational choice—the ball game or the movie?

Suppose that the probability of fine weather is $\frac{3}{5}$ and the probability that the weather will not be fine is $\frac{2}{5}$. We can now determine the expected profit for each alternative. For the ball game, the expected profit is

$$(\tfrac{3}{5})\$7000 + (\tfrac{2}{5})\$1000 = \$4600.$$

For the movie, the expected profit is

$$(\tfrac{3}{5})\$3000 + (\tfrac{2}{5})\$6000 = \$4200.$$

Thus, if the probabilities for the weather are as given above, Jones should plan for a ball game. Of course, these values are *expected* profits and might not ever actually be achieved. However, if the expected net proceeds remained the same, and if the probability of fine weather was always $\tfrac{3}{5}$, and if Jones consistently chose the ball game, then his average net proceeds should be about $4600 per year. Some years Jones will realize $7000 and some years only $1000, but, over a long period of time, the average should be close to $4600 per year.

Suppose, however, that the arrangements must be made too far in advance to get reliable probabilities for the weather. In this case, Jones can proceed to make the expected profit the same regardless of the weather. He will then leave the decision to chance; this he does by marking some slips of paper "ball game" and others "movie," mixing them in a box, and drawing one. The number of slips marked each way can be determined by equating the expected profit for the two types of weather. If the probability of drawing "ball game" is p and the probability of drawing "movie" is $1 - p$, the expected profit if the weather is fine is

$$\$7000p + \$3000(1 - p) = \$3000 + \$4000p.$$

If the weather is not fine, the expected profit is

$$\$1000p + \$6000(1 - p) = \$6000 - \$5000p.$$

The two expected profits are equal if $p = \tfrac{1}{3}$. Therefore, if Jones marks one slip "ball game" and two slips "movie" he will have the same expected profit no matter what slip is drawn and regardless of the weather. The expected profit is $4333, obtained by substituting $p = \tfrac{1}{3}$ in either formula above. Of course, the actual profit realized will never be $4333; however, if the expected net proceeds remained the same and if this procedure were followed over a number of years, the average profit should be close to $4333.

Example 2.   The following use of expectations to make decisions was employed successfully by a Dean at a certain university.

A university Dean has to obtain, long before the start of the next academic year, some idea of what his unavoidable overall expenditure will be in that next year. Given this, he can then decide what funds are free for visiting faculty, extra teaching assistants, and so on. Obtaining this information is not as easy as one might think. For example, in a particular department there may be uncertainty about four of the faculty members' plans for the next year as follows:

1. Professor A has applied to the Ford Foundation, the National Science Foundation, and one other agency for funds to spend the year in Mexico City. If he is successful, the Dean will not need to pay his salary of $10,500. If he is unsuccessful, the Dean will have to pay him the $10,500.

2. Professor B already has funds from a local research center for half the year and is attempting to find support for the other half. His salary would be $11,000 for the full year.

3. Professor C is trying to obtain a better paying position at another university. This is a secret but news has leaked out to the department chairman, who has told the Dean. Professor C would receive a salary of $9000 if he stayed next year.

4. Professor D would like to spend the second half of the year in Norway. He already has obtained support for 80% of the second half of his salary of $10,000. He is trying to decide whether he will go anyway or whether that other 20% (that is, $1000) is important to him, so that he will stay home all year and so will need to be paid.

We see that the extremes of the Dean's financial problem caused by the antics of the four professors are that he may have to provide only

$$0 + 0 + 0 + 5000 = 5000$$

(in dollars) or, on the other hand, as much as

$$10,500 + 5500 + 9000 + 10,000 = 35,000.$$

What figure does he put down? If he puts down the high figure he may have money to spare and be unable to use it properly. If he puts down the low figure he may be desperately short of funds later. How about the mean, $20,000?

In fact, the Dean worked as follows. In every case he attempted to assess a subjective probability of a professor's being at the university during the next year, by talking to the department chairman and finding out all the facts that he was able to discover. He then worked out the expected salary for each case and added the results.

In Professor A's case it seemed extremely likely that he would receive outside support. In fact, the Dean assessed the chances at about 0.95 and so calculated

$$EA = 0.95(0) + 0.05(10,500) = 525.$$

Professor B's chances of finding money for his second half year were small. The Dean assessed them at about 10%, that is, 0.10, and evaluated

$$EB = 0.10(0) + 0.90(5500) = 4950.$$

Professor C's wife was a local girl and did not want her husband to move.

Thus, although C had an even chance of getting a better offer, the Dean felt there was an 80% chance that C would eventually stay. So

$$EC = 0.20(0) + 0.80(9000) = 7200.$$

Professor D was a bachelor and had few financial commitments. The Dean felt that he would go with probability 0.90. Thus he calculated

$$ED = 0.90(10,000) + 0.10(5000) = 9500.$$

The Dean then found the sum

$$E(A + B + C + D) = 525 + 4950 + 7200 + 9500$$
$$= 22,175$$

and estimated this as his commitment for the four professors combined.

Of course, it was impossible to actually get to this figure in the event, no matter how all the questions were resolved. However, when this sort of procedure was carried out over all the departments under the Dean's control, quite reliable overall estimates emerged. Of course, much depends here on the Dean's ability to assign realistic probabilities, and it is easy to see how such a system *could* be disastrous. Nevertheless, it worked well in practice for one man at least. Deans using this method do so at their own risk!

The two examples above are both very simple ones. Other more complicated decision problems make up an extremely interesting segment of probability and statistics.

Exercises

1. A game is played as follows. Player A tosses a die; if a 1, 3, or 5 turns up, he pays player B \$1; if a 2 or a 4 turns up, B pays A \$2; if a 6 turns up, no payments are made. What is A's mathematical expectation in one toss? How much can A expect to win in 10 tosses?

2. What is the total number of heads expected in 1000 repetitions of the experiment of tossing three balanced coins?

3. What is the expected number of bases a ball player will get each time at bat if he hits singles with probability 0.2, doubles with probability 0.06, triples with probability 0.01, homers with probability 0.03, walks with probability 0.05, and otherwise is out?

4. Suppose that you are in charge of arranging for a money-making autumn bazaar and have to decide whether to hold it indoors or outdoors. If you hold it outdoors and the weather is good, you know from experience that you will net \$1000; if the weather is poor, you will net \$400. If you hold it indoors and the weather is good, you will net \$600; if the weather is poor, you will net \$800. Explain how you will decide what to do if the

probability of good weather is

   (a) $\frac{3}{5}$.   (b) unknown.

5. A motorist knows that his car needs a new tire within the next year. In his state, the police conduct random inspections and there is a chance p that his car *will* be inspected. If his car *is* inspected, he will have to obtain the tire immediately and also pay a $3 reinspection fee as well. If he can hold off buying the tire for the whole year, his money will earn an additional $1 in a savings bank. The motorist decides to wait; what has he decided about p? (The motorist's wife, when consulted, remarked: "Well, I don't know about p, but chances are you'll have a blowout, and have to pay $2000 worth of car damages, plus additional hospital expenses." The point: You may make the right decision on the problem you considered, but did you examine the right problem?)

10.4. Suggested Additional Reading

Aigner, D. J., *Principles of Statistical Decision Making*, Macmillan, New York, 1968.

Hadley, G., *Introduction to Business Statistics*, Holden-Day, San Francisco, 1968.

Neter, J., and W. Wasserman, *Fundamental Statistics for Business and Economics*, Allyn and Bacon, Boston, 3rd ed., 1966.

Sasaki, K., *Statistics for Modern Business Decision Making*, Wadsworth, Belmont, Calif., 1968.

Sielaff, T. J., and P. S. Wang, *Practical Problems in Business and Economic Statistics*, Holden-Day, San Francisco, 1968.

Answers to Exercises

Section 1.2, page 4

1. (a) random (b) deterministic (c) random (d) random (e) deterministic (f) random
4. (a) random (b) random (c) deterministic (d) random
5. all are random

Section 1.3, page 8

1. (a) $\binom{5}{k}\left(\frac{1}{2}\right)^k\left(\frac{1}{2}\right)^{5-k}$; $\frac{1}{32}, \frac{5}{32}, \frac{10}{32}, \frac{10}{32}, \frac{5}{32}, \frac{1}{32}$; 1

   (b) $\frac{1}{8}$ (c) $\binom{6}{k}\left(\frac{1}{2}\right)^k\left(\frac{1}{2}\right)^{6-k}$; $\frac{5}{16}$ (d) $\binom{100}{k}\left(\frac{1}{2}\right)^k\left(\frac{1}{2}\right)^{100-k}$

2. $1/10!$

Section 1.4, page 15

1. (a) $\frac{1}{6}$ (b) $\frac{1}{2}$ (c) $\frac{1}{3}$ (d) $\frac{2}{3}$
2. (a) $\frac{1}{4}$ (b) $\frac{1}{13}$ (c) $\frac{3}{13}$ (d) $\frac{4}{13}$
3. (a) $\frac{1}{5}$ (b) $\frac{3}{10}$ (c) $\frac{1}{2}$
4. (a) $\frac{2}{25}$ (b) $\frac{3}{50}$ (c) $\frac{1}{10}$ (d) $\frac{1}{50}$ (e) $\frac{1}{25}$
7. (a) $\frac{8}{25}$ (b) $\frac{3}{25}$ (c) $\frac{18}{25}$
8. (a) $\frac{5}{14}$ (b) $\frac{5}{36}$ (c) $\frac{3}{13}$ (d) $\frac{1}{2}$

Section 1.5, page 24

3. 90,000
5. 84

7. 30
9. 120
11. 120
13. (a) 15,600   (b) 17,576
15. 70
17. 15,120
19. (a) 26,000,000   (b) 20,700,000   (c) 22,999,977
21. (a) 48   (b) 5148   (c) 3744
23. $2^{14} \times 3^5 \times 4$
25. 27,000
27. 18
29. 11,880; 12^4
31. 1024
33. 1152; $4 \times 7!$
35. $\binom{12}{8}$

37. $\binom{52}{5}$

Section 2.1, page 30

1. (a) {i, n, c, o, v, e, t}   (b) $\{x : 1 \leq x \leq 5\}$
   (c) {Wilson, Harding, Coolidge, Hoover, Roosevelt, Truman}
   (d) {Mercury, Venus, Earth, Mars, Saturn, Jupiter, Neptune, Uranus, Pluto}
   (e) $\{x : x \text{ is an integer}\}$

Section 2.2, page 31

1. (a) 35; 7   (b) $\{x : 2 \leq x \leq 3\}$ (others are possible); $\{(100 + x)/100 : x = 1, 2, \ldots, 400\}$ (others are possible)
   (d) the null set is common

Section 2.3, page 35

1. (a) {2, 3, 4, 6, 9, 13}   (b) {9}   (c) 4   (d) 3   (e) 6   (f) {1, 2, 3, 5, 7, 8, 10, 11, 12, 14}   (g) {1, 5, 7, 8, 10, 11, 12, 14}   (h) same as (g)   (i) {1, 2, 3, 4, 5, 6, 7, 8, 10, 11, 12, 13, 14}   (j) same as (i)

Section 2.4, page 39

1. $(A - B)' = (A \cap B')' = A' \cup B$
3. $n(A - B) = 3, n(A \cap B) = 2, n(B - A) = 2, n(A \cup B)' = 4$
13. (a) 0   (b) 13   (c) 13   (d) 13   (e) 0   (f) 52   (g) 13   (h) 39   (i) 26

Section 3.1, page 43

7. {(1, 6), (2, 5), (3, 4), (4, 3), (5, 2), (6, 1)}   (b) {(5, 6), (6, 5)}   (c) union of (a) and (b)   (d) {(1, 1), (1, 2), (1, 3), (2, 1), (2, 2), (3, 1)}   (e) {(1, 1), (6, 6)}

9. $\{(4, 9), (4, W), (9, W)\}$; $\{\ \}$, $\{(4, 9)\}$, $\{(4, W)\}$, $\{(9, W)\}$, $\{(4, 9), (4, W)\}$, $\{(4, 9), (9, W)\}$, $\{(4, W), (9, W)\}$, $\{(4, 9), (4, W), (9, W)\}$
13. 32; 10; 10
15. $\{\ \}$, $\{a\}$, $\{b\}$, $\{c\}$, $\{a, b\}$, $\{a, c\}$, $\{b, c\}$, $\{a, b, c\}$

Section 3.2, page 47

1. $1/8!$
3. (a) $\frac{8}{27}$ (b) $\frac{4}{9}$ (c) $\frac{20}{27}$
5. (a) 0.6 (b) 0.1 (c) 0.6 (d) 0.9 (e) 0.6 (f) 0.4
7. (a) $\frac{3}{8}$ (b) $\frac{3}{8}$ (c) $\frac{1}{8}$
9. (a) $\frac{1}{6}$ (b) $\frac{1}{18}$ (c) $\frac{2}{9}$ (d) $\frac{1}{6}$ (e) $\frac{1}{18}$
13. sample points are not equally likely
15. $\frac{46}{135}$

Section 3.3, page 52

1. (a) $\frac{8}{13}$ (b) $\frac{5}{13}$
5. (a) $\frac{1}{4}$ (b) $\frac{1}{13}$ (c) $\frac{1}{52}$
7. $\frac{1}{6}$; $\frac{1}{3}$
9. $1 - p_1 p_2 p_3$

11. $\left[4\binom{13}{1}\binom{39}{12} - \binom{13}{1}\binom{13}{1}\binom{26}{11} - 2\binom{13}{1}\binom{13}{1}\binom{13}{1}\binom{13}{10} \right] \Big/ \binom{52}{13}$

13. (c) $\frac{2}{9}$
15. (a) 0.3 (b) 0.7

17. $\left[\binom{4}{2}\binom{48}{3} + \binom{4}{3}\binom{48}{2} + \binom{4}{4}\binom{48}{1} \right] \Big/ \binom{52}{5}$

19. $1 - \binom{N}{n} \Big/ n^N$

Section 3.4, page 57

1. (a) $\frac{1}{2}$ (b) $\frac{17}{52}$ (c) $\frac{1}{2}$
3. $\frac{1}{8}$, $\frac{1}{5}$; $1 - \frac{29}{120}$
5. $\frac{2}{3}$
7. $\frac{4}{15}$
9. (a) $\frac{3}{11}$ (b) $\frac{6}{11}$ (c) $\frac{1}{2}$
11. (a) $\frac{1}{221}$ (b) $\frac{13}{102}$ (c) $\frac{25}{204}$ (d) $\frac{1}{52}$ (e) $\frac{4}{663}$
13. (a) 0.24 (b) 0.832
15. (a) 0.12 (b) 0.78 (c) 0.4
17. (a) 0.3 (b) 0.96
19. (a) 0.2 (b) $\frac{2}{3}$ (c) 0.8
21. (a) $\frac{7}{16}$ (b) $\frac{3}{8}$ (c) $\frac{1}{4}$ (d) $\frac{2}{33}$ (e) $\frac{25}{99}$
23. $\frac{1}{7}$
25. $\frac{5}{23}$

Section 3.5, page 66

1. 0.70
3. 0.3925
5. $\frac{11}{36}$
7. (a) 0.1 (b) $\frac{1}{8}$ (c) 0.8
9. $\frac{15}{62}$
11. $\frac{3}{13}, \frac{6}{13}, \frac{4}{13}$ are the probabilities
13. $\frac{25}{61}$

Section 3.6, page 73

3. 0.8
5. (a) 0.3 (b) 0.8 (c) 0.6
7. (a) $\frac{2}{3}$ (b) $\frac{1}{4}$ (c) $\frac{11}{12}$ (d) no (e) no (f) no
11. (a) $\frac{1}{4}$ (b) $\frac{1}{2}$ (c) $\frac{3}{4}$ (d) $\frac{1}{4}$ (e) $\frac{1}{2}$
13. (a) $\frac{5}{162}$ (b) 0 (c) $\frac{85}{324}$ (d) $\frac{59}{324}$

Section 4.2, page 81

1. c, d, i, m, n, and q are continuous; the others are discrete

Section 4.3, page 85

1. no; there are an infinite number of possible values
3. sum the geometric progression
5. (a) $1/n$ (b) $1/n$ (c) $(n-3)/n$ (d) $(n-2)/n$ (e) $(n-6)/n$
7. (a) $\frac{1}{12}$ (b) $\frac{5}{12}$ (c) $\frac{11}{12}$ (d) 0

Section 4.4, page 96

1. $9/10^9$
3. $24/5^6$
5. (a) $\frac{3}{8}$ (b) $\frac{11}{16}$ (c) $\frac{11}{16}$
7. (a) $189/4^7$ (b) $(\frac{7}{13})(\frac{12}{13})^6$ (c) $35 \times 81,000/13^7$ (d) $\frac{7}{128}$
9. (a) $\frac{135}{4096}$ (b) $\frac{15}{64}$
11. (a) $\frac{1}{243}$ (b) $\frac{40}{243}$ (c) $\frac{32}{243}$
13. (a) $(\frac{2}{3})^8$ (b) $1120/3^8$ (c) $1024/3^7$ (d) $16/3^8$
15. (a) $35/2^7$ (b) $(\frac{1}{2})^7$ (c) $(\frac{1}{2})^7$ (d) $(\frac{1}{2})^7$
17. (a) $\frac{63}{512}$ (b) $78,125/1,679,616$ (c) $\frac{19}{256}$
19. $\frac{3}{8}$
21. (a) $\frac{125}{3888}$ (b) $\frac{23}{648}$ (c) $\frac{1}{81}$
23. the probability of $(n-x)$ failures equals the probability of x successes
25. (a) 0.121 (b) 0.367 (c) 0.259

Section 5.1, page 107

1. (a) $F(x) = \sum_{t=0}^{x} \binom{10}{t} (0.8)^t (0.2)^{10-t}$ (b) $F(x) = \sum_{t=0}^{x} \binom{8}{t} \left(\frac{2}{3}\right)^t \left(\frac{1}{3}\right)^{8-t}$

3. (a) $F(x) = \sum_{t=1}^{x} (1-p)^{t-1}p = 1 - (1-p)^x$

   (b) $F(x) = \sum_{t=r}^{x} \binom{t-1}{r-1} p^r(1-p)^{t-r}$

   (c) $F(x) = \sum_{t=0}^{x} \binom{r+t-1}{t} p^r(1-p)^t$

5. (b) $(2x+1)/63$

7. $(1-p)^b$

Section 5.2, page 110

1. 0.6
3. 0.4
5. (a) $\frac{2}{3}$ (b) $\frac{1}{3}$ (c) 0
7. $\frac{15}{88}$

Section 5.3, page 120

1. $5; \frac{149}{5}$
3. (a) $15; 9$ (b) $11; 7$ (c) $\frac{1341}{5}; 18$ (d) $\frac{596}{5}; 8$
5. $\frac{19}{3}; 3\sqrt{5}$
7. $3.5; \frac{35}{12}$
9. $p; \sqrt{p(1-p)}$
11. $3; \sqrt{1.5}$
13. (a) $37; \sqrt{\frac{74}{3}}$ (b) $3; \sqrt{\frac{108}{37}}$
15. $\frac{16}{3}; \frac{16}{9}$
21. expand $(x-\mu)^2$
23. expand $(x-\mu)^3$
25. expand $(x-\mu)^4$
27. $n(n-1)(n-2)(n-3)p^4$

Section 5.4, page 124

1. 0.027
3. 0.219

Section 6.1, page 130

1. $\frac{134,976}{997,513}$
3. (a) $\frac{3}{4025}$ (b) $\frac{6}{4025}$
5. (a) $\binom{4}{4}\binom{48}{1} / \binom{52}{5} = \frac{1}{2,869,685}$ (b) $\binom{4}{1}\binom{48}{4} / \binom{52}{5} = \frac{3243}{10,829}$

   (c) $\binom{4}{0}\binom{48}{5} / \binom{52}{5} = \frac{35,673}{54,145}$

7. $\left[\binom{6}{1}\binom{4}{2} + \binom{6}{2}\binom{4}{1} + \binom{6}{3}\binom{4}{0}\right] / \binom{10}{3} = \frac{29}{30}$

9. $\binom{4}{2}12\binom{4}{2}11\binom{4}{1}\Big/\binom{52}{5}$

Section 6.1, page 131

1. $\left[\binom{40}{0}\binom{20}{10} + \binom{40}{1}\binom{20}{9} + \binom{40}{2}\binom{20}{8} + \binom{40}{3}\binom{20}{7}\right]\Big/\binom{60}{10}$

3. $1 - \left[\binom{6}{0}\binom{24}{6} + \binom{6}{1}\binom{24}{5}\right]\Big/\binom{30}{6}$

5. $\left[\binom{6}{3}\binom{14}{2} + \binom{6}{4}\binom{14}{1} + \binom{6}{5}\binom{14}{0}\right]\Big/\binom{20}{5}$

Section 6.1, page 134

1. $\frac{20}{3}$; $\frac{1000}{531}$
3. $\frac{6}{5}$; $\frac{576}{725}$
5. $\frac{3}{2}$; $\frac{63}{76}$

Section 6.2, page 142

1. (a) 0.125   (b) 0.049   (c) $e^{-20}(20)^{20}/20!$ (d) 0.0156
3. (a) 0.214   (b) 0.067   (c) 0.082   (d) 0.456
5. e^{-8} (b) 0.092   (c) $1 - e^{-8}$
7. 0.981

Section 6.2, page 144

1. 0.035377; 0.03609
3. 0.0002

Section 7.1, page 151

1. no
5. (a) 0   (b) $\frac{1}{9}$ (c) $\frac{4}{9}$ (d) $\frac{4}{9}$ (e) $\frac{1}{3}$ (f) $\frac{1}{3}$ (g) 1   (h) 1
7. (a) 0   (b) $\frac{1}{6}$ (c) $\frac{1}{4}$ (d) $\frac{1}{3}$ (e) $\frac{1}{3}$ (f) $\frac{1}{3}$ (g) 1   (h) 1
9. (a) joint probabilities are $\frac{1}{36}$; marginal probabilities are $\frac{1}{6}$
   (b) yes

Section 7.1, page 155

3. $(2x^2 - 3)/20$; $(13 - 2y)/20$; $p(x|y) = (x^2 - y)/(13 - 2y)$; $q(y|x) = (x^2 - y)/(2x^2 - 3)$

Section 7.3, page 160

1. (a) $\frac{1}{2}$ (b) 1   (c) $\frac{1}{4}$ (d) $\frac{1}{2}$ (e) 0
3. (a) $\frac{5}{2}$ (b) $\frac{11}{6}$ (c) $\frac{1}{2}$ (d) $\frac{17}{36}$ (e) 0
5. (a) 0   (b) 2   (c) 1   (d) $\frac{1}{4}$ (e) 0
7. (a) $\frac{17}{11}$ (b) $\frac{70}{33}$ (c) $\frac{30}{121}$ (d) $\frac{710}{1089}$ (e) $-\frac{6}{1331}$

Section 7.3, page 164

1. 0
3. 0
5. $11\sqrt{5}/70$
7. $\sqrt{33}/99$

Section 7.4, page 167

1. (a) $\frac{5}{3888}$ (b) $\frac{175}{10,368}$
3. $\frac{15}{256}$

5. (a)

(x_1, x_2, x_3)	Probability
$(0, 0, 4)$	$\frac{16}{625}$
$(0, 1, 3)$	$\frac{48}{625}$
$(1, 0, 3)$	$\frac{48}{625}$
$(0, 2, 2)$	$\frac{54}{625}$
$(2, 0, 2)$	$\frac{54}{625}$
$(1, 1, 2)$	$\frac{108}{625}$
$(0, 3, 1)$	$\frac{27}{625}$
$(3, 0, 1)$	$\frac{27}{625}$
$(2, 1, 1)$	$\frac{81}{625}$
$(1, 2, 1)$	$\frac{81}{625}$
$(1, 3, 0)$	$\frac{81}{2500}$
$(3, 1, 0)$	$\frac{81}{2500}$
$(2, 2, 0)$	$\frac{243}{5000}$
$(4, 0, 0)$	$\frac{81}{10,000}$
$(0, 4, 0)$	$\frac{81}{10,000}$

(b) $p(0) = q(0) = 0.2401, r(0) = 0.1296$
$p(1) = q(1) = 0.4116, r(1) = 0.3456$
$p(2) = q(2) = 0.2646, r(2) = 0.3456$
$p(3) = q(3) = 0.0756, r(3) = 0.1536$
$p(4) = q(4) = 0.0081, r(4) = 0.0256$

(c) $p(0|0) = \frac{1}{16}, q(0|2) = \frac{16}{49}, r(0|1) = \frac{27}{343}$
$p(1|0) = \frac{1}{4}, \quad q(1|2) = \frac{24}{49}, r(1|1) = \frac{108}{343}$
$p(2|0) = \frac{3}{8}, \quad q(2|2) = \frac{9}{49}, r(2|1) = \frac{144}{343}$
$p(3|0) = \frac{1}{4} \qquad\qquad\qquad r(3|1) = \frac{64}{343}$
$p(4|0) = \frac{1}{16}$

(d) $-\frac{3}{7}, -\sqrt{\frac{2}{7}}, -\sqrt{\frac{2}{7}}$

7. (a) $7(9)^6/8(10)^9$ (b) $7(9)^8/32(10)^9$
9. $\frac{31}{81}$

13. (a) $\dfrac{8!}{2!2!1!3!}\left(\dfrac{5}{18}\right)^2\left(\dfrac{2}{9}\right)^2\left(\dfrac{1}{9}\right)\left(\dfrac{7}{18}\right)^3$ (b) $\binom{5}{2}\binom{4}{2}\binom{2}{1}\binom{7}{3}\bigg/\binom{18}{8}$

Section 8.1, page 172

1. $p(0) = \frac{1}{9}, p(1) = \frac{2}{9}, p(2) = \frac{1}{3}, p(3) = \frac{2}{9}, p(4) = \frac{1}{9}$ (b) same as (a) (c) same as (a) (d) $p(0) = \frac{1}{27}, p(1) = \frac{1}{9}, p(2) = \frac{2}{9}, p(3) = \frac{7}{27}, p(4) = \frac{2}{9}, p(5) = \frac{1}{9}, p(6) = \frac{1}{27}$

3.

$x + y$	-7	-6	-5	-4	-3	-2	-1	0	1	2	3	4	5	6	7
$p(x + y)$	$\frac{1}{320}$	$\frac{1}{160}$	$\frac{1}{64}$	$\frac{1}{40}$	$\frac{1}{20}$	$\frac{1}{20}$	$\frac{1}{10}$	$\frac{1}{10}$	$\frac{51}{320}$	$\frac{19}{160}$	$\frac{59}{320}$	$\frac{1}{20}$	$\frac{3}{40}$	$\frac{1}{40}$	$\frac{3}{80}$

Section 8.2, page 176

1. $k/2$; $k/4$; can also use binomial with $n = k$, $p = q = \frac{1}{2}$

3. (a) $\sum\limits_{i=1}^{n} \lambda_i$ (b) $p(\bar{x}_n) = e^{-\lambda} \lambda^{n\bar{x}_n}/(n\bar{x}_n)!$, $\bar{x}_n = 0, 1, \ldots$, where $\lambda = \sum\limits_{i=1}^{n} \lambda_i$

5. n and $(p_1 + p_2 + p_3)$

9. $a_1^2 V(X_1) + a_2^2 V(X_2) + 2a_1 a_2 \operatorname{Cov}(X_1, X_2)$; $\sum a_i^2 V(X_i) + \sum\sum a_i a_j \operatorname{Cov}(X_i, X_j)$

11. $-\frac{1}{4}$

Section 8.3, page 179

1. (a) $p(-1) = p(1) = \frac{1}{4}$, $p(0) = \frac{1}{2}$ (b) $p(-1) = p(1) = \frac{1}{16}$, $p(-\frac{1}{2}) = p(\frac{1}{2}) = \frac{1}{4}$, $p(0) = \frac{3}{8}$ (c) $p(-1) = p(1) = \frac{1}{256}$, $p(-\frac{3}{4}) = p(\frac{3}{4}) = \frac{1}{32}$, $p(-\frac{1}{2}) = p(\frac{1}{2}) = \frac{7}{64}$, $p(-\frac{1}{4}) = p(\frac{1}{4}) = \frac{7}{32}$, $p(0) = \frac{35}{128}$ (d) $p(-1) = p(1) = 1/2^{16}$, $p(-\frac{7}{8}) = p(\frac{7}{8}) = 1/2^{12}$, $p(-\frac{3}{4}) = p(\frac{3}{4}) = 15/2^{13}$, $p(-\frac{5}{8}) = p(\frac{5}{8}) = 35/2^{12}$, $p(-\frac{1}{2}) = p(\frac{1}{2}) = 455/2^{14}$, $p(-\frac{3}{8}) = p(\frac{3}{8}) = 273/2^{12}$, $p(-\frac{1}{4}) = p(\frac{1}{4}) = 1001/2^{13}$, $p(-\frac{1}{8}) = p(\frac{1}{8}) = 715/2^{12}$, $p(0) = 6435/2^{15}$

3. (a) $p(-1) = p(1) = \frac{1}{9}$, $p(-\frac{1}{2}) = p(\frac{1}{2}) = \frac{2}{9}$, $p(0) = \frac{1}{3}$ (b) $p(-1) = p(1) = 1/3^4$, $p(-\frac{3}{4}) = p(\frac{3}{4}) = 4/3^4$, $p(-\frac{1}{2}) = p(\frac{1}{2}) = 10/3^4$, $p(-\frac{1}{4}) = p(\frac{1}{4}) = 16/3^4$, $p(0) = 19/3^4$ (c) $p(-1) = p(1) = 1/3^8$, $p(-\frac{7}{8}) = p(\frac{7}{8}) = 8/3^8$, $p(-\frac{3}{4}) = p(\frac{3}{4}) = 4/3^6$, $p(-\frac{5}{8}) = p(\frac{5}{8}) = 112/3^8$, $p(-\frac{1}{2}) = p(\frac{1}{2}) = 266/3^8$, $p(-\frac{3}{8}) = p(\frac{3}{8}) = 56/3^6$, $p(-\frac{1}{4}) = p(\frac{1}{4}) = 784/3^8$, $p(-\frac{1}{8}) = p(\frac{1}{8}) = 1016/3^8$, $p(0) = 41/3^5$ (d) $p(-1) = p(1) = 1/3^{16}$, $p(-\frac{15}{16}) = p(\frac{15}{16}) = 16/3^{16}$, $p(-\frac{7}{8}) = p(\frac{7}{8}) = 136/3^{16}$, $p(-\frac{13}{16}) = p(\frac{13}{16}) = 800/3^{16}$, $p(-\frac{3}{4}) = p(\frac{3}{4}) = 3620/3^{16}$, $p(-\frac{11}{16}) = p(\frac{11}{16}) = 13{,}328/3^{16}$, $p(-\frac{5}{8}) = p(\frac{5}{8}) = 41{,}328/3^{16}$, $p(-\frac{9}{16}) = p(\frac{9}{16}) = 12{,}272/3^{14}$, $p(-\frac{1}{2}) = p(\frac{1}{2}) = 28{,}730/3^{14}$, $p(-\frac{7}{16}) = p(\frac{7}{16}) = 178{,}880/3^{15}$, $p(-\frac{3}{8}) = p(\frac{3}{8}) = 332{,}072/3^{15}$, $p(-\frac{5}{16}) = p(\frac{5}{16}) = 555{,}152/3^{15}$, $p(-\frac{1}{4}) = p(\frac{1}{4}) = 840{,}112/3^{15}$, $p(-\frac{3}{16}) = p(\frac{3}{16}) = 1{,}155{,}280/3^{15}$, $p(-\frac{1}{8}) = p(\frac{1}{8}) = 1{,}447{,}720/3^{15}$, $p(-\frac{1}{16}) = p(\frac{1}{16}) = 4{,}929{,}152/3^{16}$, $p(0) = 577{,}403/3^{14}$

Section 9.3, page 189

3. (d) equal probabilities for all states

5. initial state

7. (b) $P_4 = \begin{bmatrix} 0 & 1 & 0 & 0 \\ 0 & 1 & 0 & 0 \\ 0 & 1 & 0 & 0 \\ 0 & 1 & 0 & 0 \end{bmatrix}$

9. (b) $P_4 = \begin{bmatrix} 1 & 0 & 0 & 0 \\ 1 & 0 & 0 & 0 \\ \frac{29}{32} & \frac{1}{32} & \frac{1}{32} & \frac{1}{32} \\ \frac{29}{32} & \frac{1}{32} & \frac{1}{32} & \frac{1}{32} \end{bmatrix}$

11. (b) $P_4 = \begin{bmatrix} 0 & 0 & 0 & 0 \\ \frac{153}{256} & \frac{1}{256} & \frac{51}{256} & \frac{51}{256} \\ 0 & 0 & \frac{1}{2} & \frac{1}{2} \\ 0 & 0 & \frac{1}{2} & \frac{1}{2} \end{bmatrix}$

Section 9.4, page 200

1. (a) q (c) $1 - 2q$; 1

Section 9.5, page 207

1. $p(-5) = p(5) = \frac{1}{32}, p(-3) = p(3) = \frac{5}{32}, p(-1) = p(1) = \frac{10}{32}; p(0) = p(1) = \frac{10}{32}; p(2) = p(3) = \frac{5}{32}, p(4) = p(5) = \frac{1}{32}; p(0) = \frac{3}{8}, p(2) = \frac{1}{4}, p(4) = \frac{3}{8}$
3. $r = 1: \frac{1}{2}, 0, \frac{1}{8}, 0, \frac{1}{16}; r = 2: 0, \frac{1}{4}, 0, \frac{1}{8}, 0$

Section 10.2, page 212

1. (a) 0 (b) 0
3. $\frac{1}{2}$ cent gain
5. $2.50; $7.50; $0.75

Section 10.3, page 216

1. $1/6; $10/6
3. 0.52
5. $p < \frac{1}{4}$

Index